LOOKING BEYOND THE HIGHWAY

LOOKING BEYOND THE HIGHWAY

DIXIE ROADS AND CULTURE

EDITED BY CLAUDETTE STAGER AND MARTHA CARVER

THE UNIVERSITY OF TENNESSEE PRESS / KNOXVILLE

First Edition.

This book is printed on acid-free paper.

Library of Congress Cataloging-in-Publication Data

Looking beyond the highway : Dixie roads and culture / edited by
Claudette Stager and Martha Carver.— 1st ed.

p. cm.

Includes index.

ISBN 1-57233-467-3 (hardcover : alk. paper)

1. Roadside architecture—Southern States.
2. Southern States—Description and travel.
3. Southern States—Social life and customs.
4. Dixie Highway—History.
5. Dixie Highway—Description and travel.

I. Stager, Claudette.
II. Carver, Martha, 1954–

F216.2.L66 2006
975'.042—dc22 2005028531

CONTENTS

ILLUSTRATIONS

FIGURES

TABLES

ACKNOWLEDGMENTS

The editors would like to extend a special thanks to Beth Savage with the National Park Service and Tammy Allison with the Tennessee Department of Transportation for their editorial comments and assistance. Kristen Broussard has our gratitude for putting finishing touches on the manuscript.

INTRODUCTION

CLAUDETTE STAGER

Looking beyond the Highway: Dixie Roads and Culture is an outgrowth of work fostered by the Society for Commercial Archeology (SCA) that primarily deals with southern roads and southern roadside culture. The SCA began in 1977 and is the oldest national organization that encourages the study of the unique historic significance of twentieth-century commercial landscape and culture. Thirteen years after its founding, SCA looked at the impact of roads on culture with the publication of *Roadside America: The Automobile in Design and Culture,* edited by Jan Jennings (Iowa State UP, 1990). In the years between the publication of the first book and the work on this second book, interest in the early-twentieth-century roadside has grown tremendously. Books and organizations on specific topics, such as Route 66 or the National Road, have been published, and organizations with resource-specific focuses, such as diners, theaters, or the Lincoln Highway, have become more popular. However, few of these books or organizations emphasized the South and its role in roadside history and culture, automobile tourism, and the development of the system of roads we travel daily.

Today it seems natural to want to drive south to warmer climates for a vacation, but in the early twentieth century there was no easy way to head south. There were no federal-government-sponsored highways to Florida, Tennessee, or the Carolinas. It was not until the 1920s that the southern state highway departments took a lead role in road construction and not until the late 1920s that a federal routing plan systematically linked roads from state to state. Travelers who wanted to experience the warmer climate, cooling waters, or fresh air of the mountains had to find a route south before they could even begin the trip. Few people had the time or finances for such journeys. Those who could were twentieth-century pioneers, taking several days to motor to their destination while hoping that the roads would be passable and that they might find somewhere along the road to camp. The Dixie Highway was one of the first comprehensive interstate routes to bring people south. Begun by northern entrepreneur Carl Fisher, the road was to be a major route to the South so that people would visit his Florida hotels.

Roadway pioneers such as Fisher, who were concerned with profit, or organizations such as the Daughters of the American Revolution, whose primary interest

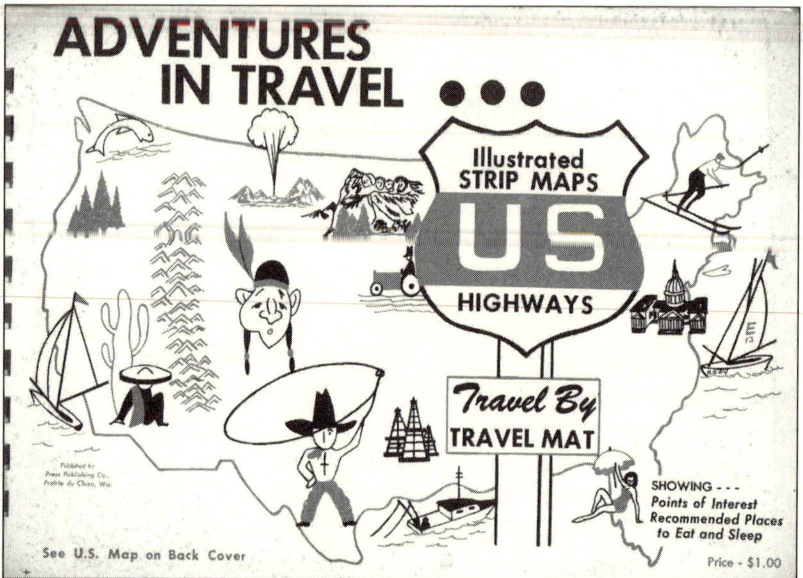

FIG. 0.1. Until the modern interstate system was in place, many journeys were "Adventures in Travel," as shown by this set of placemats displaying U.S. highways. From the author's collection.

was in marking approximations of a historic route such as the Natchez Trace, thought they were helping communities along the new highway routes. However, problems in southern road development quickly manifested themselves in the form of opposition to the road or a desire for a different course for the road. With no official government status, these early efforts at establishing interstate roads into the South required perseverance and determination. A prime example of this occurred while determining the route of the Dixie Highway, nicknamed the "Second Battle of Chattanooga."

Settling on the route was only the beginning of the road development. Early automobile travelers to the South might drive on dirt, concrete, macadam, or brick roads before getting to their destination. Just as in laying out the overall route, local interests played an important part in determining just what "their" part of the road would look like. If people were going to drive several hundred miles to southern vacation destinations, even on improved roads, there needed to be places to stop for brief respites from the road, places to eat or to sleep. As a result hotels, and later motels, were built to serve the transitory motorist. In addition business owners, town boosters, and local governments wanted the traveler to stay and visit their communities, so tourism-related enterprises proliferated near hotels and motels. In fact, as motoring became easier and cheaper, motels built right next to the road

FIG. 0.2. The *Highways into the Future* publication provided the latest on building and using roads. From the author's collection.

overshadowed urban hotels. Even adding parking garages near urban hotels could not halt the popularity of motels.

Essentially this anthology illustrates that early-twentieth-century interstate roads and the resulting roadside culture played a significant role in making southern states accessible. There is no doubt that these roads opened up the South, but some scholars argue that roads constructed for the farm-to-market economy were more significant. In reality both developments were important and both affected the South. The farm-to-market movement revolved around the trade in goods and getting them in and out of the South as quickly as possible. The roadside culture of the southern states involved having people actually experience the South and all that it had to offer. The road also has had a more permanent impact on the South because of the associated built environment. Much of this built environment, from motels to drive-ins to eateries, has disappeared.

Occasionally the popular media will do a story on how homogenous the United States has become. This is usually blamed on chain businesses and our current interstate system. At the same time, there are still are many misconceptions about the South as it was historically and as it is today. With fast roads and short vacation times, travelers speed through a state such as Tennessee, stopping only at an interstate rest area, and claim they have been there. There is a southern culture, and the chapters in this anthology reveal a small portion of it and how the early interstate roads influenced some of the development of the South. The writing in *Looking beyond the Highway: Dixie Roads and Culture* represents some of the best of southern culture.

Looking beyond the Highway: Dixie Roads and Culture can generally be divided into two areas of interest—the development of roads and the resources that have strong connections to the roads. The term "generally" is used because all of these chapters tie in together and the basic premises overlap. There are actually several "subthemes" that can be found in the anthology. Chapters on roads, nature and the impact on tourism, tourism and how it impacted nature, accommodations, places to stop along the road, destinations, and even a touch of religion are in the book. The variety and scope of all these ideas is shown in the two final chapters, in which the authors write about the impact of the road and roadside on culture. Whether the roads were designed to bring northern vacationers to the South or to open up a particular part of a state for development, they did not develop in a vacuum with no impact on the land and culture. The term "roadside" is often used to describe the built environment that followed the highway, but this is almost a misnomer. The cultural and environmental impact of the road extended far beyond the highway.

The Dixie Highway was not the only road built to bring more tourists and tourist dollars to the South, but it was arguably the route that had the largest impact on the development of southern states. No writing on the Dixie Highway can omit mention of Carl Fisher, who was the foremost promoter behind the north-south route. In "The Best Road South: The Failure of the Dixie Highway in Indiana," Suzanne Fischer explains how the highway's path was chosen in Fisher's home state. The mountains and swamps of the southern states had been road-building obstacles for years, resulting in few major routes, but a relatively flat northern state such

FIG. 0.3. *The Motorist's Luncheon Book* and the Touring card game reflect the fact that there was little for motoring travelers to do and few places to stop when heading out on a long trip. From the author's collection.

as Indiana already had a network of highways. The failure to acknowledge the established routes—they were used because of their convenient location—and reuse them for the new Dixie Highway resulted in the failure of the highway as a southern transportation route in Indiana.

The Dixie Highway Association began meeting in Chattanooga, Tennessee, in 1915. Martha Carver presents an overview of the development of the association and the road itself in her chapter "Drivin' the Dixie Highway in Tennessee." Even as the home state of the Dixie, Tennessee had problems in setting the route and maintaining the road. Counties vied for the right to have the new Dixie Highway come through their communities. Focusing on the development of the road in Tennessee, Carver explains the two routes of the Dixie Highway through Tennessee and closes with the end of the Dixie Highway Association.

Jeffrey L. Durbin details the highway's history in Georgia in "Heading South without Getting Sidetracked: The Dixie Highway in Georgia." As was the case in Tennessee, there were multiple routes, including one known as the Battlefield Route because of its proximity to battles fought during the Civil War in 1864. Out of nearly six thousand miles of Dixie Highway, Georgia had nearly fifteen hundred miles traversing the state's varied terrain. Durbin notes that while officials did what they could to espouse the road, private entrepreneurs sprung up along the highway and did their own promoting.

Instead of focusing on the political nature of building the Dixie Highway, Walter S. Marder discusses the importance of choosing the right building material. "Pleasing the Eye: Brick Paving and the Dixie Highway in the Sunshine State" presents a brief history of brick paving for roads and highways, concentrating on

Florida's choice of brick for parts of the Dixie Highway. Bricks were first used for road paving in Ohio in 1893, and the state of Florida paved its first brick highway around 1912. Florida was the terminus of the Dixie Highway, and its brick-clad road would provide the last impression of the road a tourist would have when arriving from the northern states. Marder tells how the paving material was chosen and why it was finally discarded for this use in Florida.

Once travelers reached Florida, they might drive on the Tamiami Trail, which spanned the state from Tampa on the west coast to Miami on the east coast. Remember that Miami was the bailiwick of Carl Fisher, so while not all roads would lead to Miami, many eventually did. In "Connecting the East and West Coasts: The Taimiami Trail of the Sunshine State," Carrie Scupholm discusses its development. Even though part of the route overlapped the Dixie Highway, it was always known locally as the Tamiami Trail.

In Arkansas the establishment of roads was controlled through local "road improvement districts." Christie H. McLaren's "The Dollarway Road and Road Improvement Districts in Arkansas, 1913–1921" discusses how the improvement districts hindered and helped Arkansas's system of roads. Ultimately problems and scandals with the road improvement districts in Arkansas were known nationwide. The Dollarway Road, the state's first paved road, is presented as an example of how a road could be built.

The Natchez Trace Parkway is both a southern road and a destination in and of itself. Begun in 1938 and completed in 2005, the parkway is an example of the federal government's efforts to commemorate an earlier road—the 1801 Natchez Trace. The road is a limited access parkway that functions as an interstate route and as a linear park for its 444-mile length through three states. The Natchez Trace Parkway, like the original trace, traverses a route from Natchez, Mississippi, to Nashville, Tennessee. In "The Daughters of the American Revolution, Roane F. Byrnes, and the Birth of the Natchez Trace Parkway," Sara Amy Leach explains how southern women, especially Roane Fleming Byrnes and the Daughters of the American Revolution, spearheaded this movement. It was a mix of private and public initiatives that resulted in a new road commemorating the old one, although the new road was built for automobiles and tourism, not military needs.

Automobile tourists had to have places to stay during their drive and at their destinations. Many scholars have studied and written about auto camping and motor courts, but few have looked at urban accommodations for tourists. R. Stephen Sennott looks at this phenomenon in "Roadside Luxury: Urban Hotels and Modern Streets along the Dixie Highway." Hotels had generally been constructed and operated with the traveling salesman in mind. Adapting this building form to families traveling long distances in cars meant adding amenities for people and for automobiles. Garages were built in conjunction with urban hotels. At the same time, the City Beautiful movement flourished, resulting in improvements to the overall planning of cities, including the new urban hotels for tourists.

Roadside motels and hotels might differ in design from established urban ones by having more amenities, such as porches and balconies, easy access to the ocean, or restaurants on site. These features were the keys to attracting visitors who were traveling to a beach or to a mountain resort. With the opening of the Great Smoky Mountains National Park in the early twentieth century, more people had the chance to experience summer in the mountains. Blythe Semmer's "Tourist Lodging in the Great Smoky Mountains and the Transition from Regional to National Style" shows how tourist accommodations adapted to the exquisite natural beauty of the mountains.

Myrtle Beach hotels and motels developed in yet another regional manner. In "Myrtle Beach: Music and Motels" Katherine Fuller describes the changing architectural designs of the beach resort's motels. However, as much as the accommodations, it was the beach music, based on African American rhythm and blues music that eventually made the resort popular.

In an era when air conditioning was a rarity, caves provided a cool break for the traveler. Ruth Nichols Kennoy and Robbie D. Jones write about caves as tourist attraction in "Caving and Clogging: Keepin' Cool in Tennessee Caves." Major caves such as Mammoth Cave in Kentucky were opened in the nineteenth century, but it was the popularity of automobile travel that resulted in increased commercial use of caves in the early twentieth century. Dances and tours were popular activities in southern caves until the interstate system rerouted tourists in the 1950s. Caves with stable year-round temperatures are found throughout the country, but they have a particularly important place in southern culture.

Why should people get dressed for church on vacation when they could come as they were, thus taking advantage of two aspects of roadside culture, the automobile and the drive-in theater? Carrie Scupholm explores a phenomenon that could fully develop only in warmer climates such as Florida in "Park 'n' Pray: An Examination of Drive-in Religion in Florida." With the prosperity of post–World War II United States, families were buying cars and driving everywhere—to eat, shop, see movies, and vacation. Drive-in movie theaters, popular evening entertainment spots, were deserted spaces during the daylight hours. That is, until someone considered using them for church services on Sunday mornings. Eventually Florida churches built drive-ins not for movies, but solely for outdoor church services. This allowed families to "come as you are" and solved the problem of how to accommodate vacationers who wanted to attend church services while away from home.

Automobile travel generated an increase in advertising along travel routes. Considered eyesores by some and diversions by others, billboards could range from advertising mom-and-pop businesses to the well-known BURMA SHAVE and SEE ROCK CITY signs. One singular advertising effort, not only intended for long-distance travelers but also for local residents, took advantage of the roadside. Concrete crosses and other religious markers along what were major roads in the early to mid–twentieth century, began as religious symbolism but have become roadside icons today. Martha Carver's "Get Right with God: Harrison Mayes's Roadside Advertising Campaign for the Lord" shows how these religious symbols along well-traveled routes evolved from the evangelical message that Mayes had intended into roadside folk art.

Karl Puljak's chapter looks at a building form that most have seen but have given little thought to—the fireworks stand. From the temporary shed erected only around holidays to the neon-lit facades glaring out at night along southern highways and interstates, the fireworks stand is ubiquitous. "Buy One, Get Two Free: Fireworks" discusses why the stands are located where they are and mentions laws involved with selling fireworks. Finally, Puljak gives us a typology of this unusual building type.

Although fireworks stands are everywhere, another roadside property type, observation towers, are rarely seen. Whether this has to do with safety reasons or topography, few of these towers exist. There are several in Florida as Kimberly Hinder shows in "Viewpoints and Points of View: Florida's Early Tourist Towers." Looking out over the surrounding landscape is a vacation favorite; where there were no natural features to climb, entrepreneurs built towers to entice the weary traveler to stop. The Dixie Highway in Florida had several towers that are discussed in this chapter.

Two authors use two quite different approaches to summarize roadside properties and how automobile tourism affected them. Kevin J. Patrick writes about the dual nature of vacationing and the vacation roadside landscape in "Seeing the Scenic Upland South: Mother Nature and the Morphology of Tourist Landscapes." Families travel on vacation in order to relax and get away from everyday routines. At the same time there is a great desire to do something new and not waste a minute of the time away from work or school. Many scenic attractions reflect this duality. The scenic beauty of the Natural Bridge in Virginia, Mammoth Cave in Kentucky, and Ruby Falls in Tennessee are all part of or adjacent to man-made attractions built to intrigue the tourist.

Finally, Robert M. Craig relates automobile travelers and locales to the pilgrims and their stops in Chaucer's *Canterbury Tales*. In a lighthearted comparison between the centuries, titled "Pilgrimage Route to Paradise: The Sacred and Profane along the Dixie Highway," he looks at advertising and souvenirs, accommodations and "tourist traps." Chaucer's pilgrims may have walked to shrines while the twentieth-century pilgrim drove along the Dixie Highway, but their experiences were similar.

Craig looks at tourists heading south as the worshipers of vacations and sunshine, a new type of religion for a mobile society.

Today's modern system of roads and the U.S. interstate system are controversial in planning, environmental, and historic preservation fields, to say nothing of concerns among the public. Many will argue that roads have desecrated the landscape of the country and led to the destruction of cities. Nevertheless our way of life would be quite different from what it is today without the interstate roads developed in the early twentieth century. We may complain, but we also use these roads daily. We may grouse about fast food and gas stations at every intersection, but we stop there, not at the mom-and-pop stores further down the road, because of convenience. One of the aims of this anthology is to get the reader off those main roads and back onto the roads less traveled.

The editors hope that after reading these chapters, the reader will look past the highway and discover the impact that the road has had on culture and the built environment in the South and in their neighborhood.

CHAPTER
1

THE BEST ROAD SOUTH: THE FAILURE OF THE DIXIE HIGHWAY IN INDIANA

SUZANNE FISCHER

In some ways, Indiana, the home state of Carl Fisher, "the Father of the Dixie Highway," represented the conflicting nature of the effort to get the Dixie Highway built. Fisher, founder of the Prest-O-Lite Company that made carbide headlights and developer of the Indianapolis Motor Speedway, was an avid Good Roads enthusiast and an early leader in the Lincoln Highway Association. Investing in Florida, he set about to create a good north-south road to enable northerners to travel easily to the South. Fisher philosophically believed that the route—a point on which he and Henry Joy had strongly disagreed in regards to the Lincoln Highway—should not necessarily be the straightest route between two points but rather be a route based on support from local towns and businesses. However, unlike the poor road systems in the more southern states where the Dixie was built, Indiana already had a system of operable roads. As a result the indirect routing decision influenced the development and the history of the Dixie Highway in its founder's home state quite differently than intended.

By the late 1910s, when the automobile became available to average citizens and not merely the wealthy few, vacationing took on a new form and frequency. Some Hoosiers liked to vacation in the South, to escape cold Indiana winters and to

FIG. 1.1. Dixie Highway promoter Carl Fisher (on left, with boater hat) stands one block north of the intersection of the highway with the National Road in Indianapolis. Charles W. Fairbanks, former vice president, is driving. Photo courtesy of the Indiana Historical Society Library, Bass Photo Company Collection, negative number C-513.

attend the popular auto races at Daytona Beach. One of their future vacation destinations was to be Miami, developed into a resort area in the early twentieth century by Indianapolis entrepreneur Carl Fisher (fig. 1.1). But, before Hoosiers—or anyone else—could escape to Miami by car, they had to have good roads that were passable in both winter and summer. In 1914 Fisher advocated the construction of a highway connecting north and south, similar to his east-west Lincoln Highway project. His original name for the new route was the "Hoosier Land-to-Dixie Highway," later shortened to just "Dixie Highway.[1]

Promoters first conceived the Dixie Highway in 1914 as a north-south tributary to the transcontinental Lincoln Highway, still under construction at that time. The idea of an automobile road uniting north and south appealed to many travelers and businessmen in the early twentieth century, but historians credit the first plans for such a route to Hoosier Motor Club president W. S. Gilbreath. According to Howard Lawrence Preston in *Dirt Roads to Dixie,* Fisher let Gilbreath take the credit for the Dixie Highway, just as he let Henry Joy take credit for the Lincoln Highway. Fisher liked to start projects but lost interest once work began.

At least in the beginning, the actual strategy and planning of the Dixie Highway were carried out by Indiana politicians and businessmen. Already crossed by the Lincoln Highway and the old National Road, Indiana was bound to capture a great deal of commercial and tourist traffic with the addition of a north-south highway—

a useful boost to the state's economy. The rest of the country would benefit from the proposed highway as well. Besides making travel easier for auto tourists, a through road would enable farmers to ship produce quickly by motor truck rather than take the chance of it spoiling while waiting on a railway siding.

Much like its predecessor the Lincoln Highway, the Dixie Highway was not a new *road* but a new *route* cobbled together from a network of existing roads. Its promoters planned to make the Dixie Highway the best motor route in the country. Fisher, Gilbreath, and other businessmen promoting the road decided that the route was to be profusely marked with signs so travelers would not get lost. Construction crews would improve sections where the existing roadbed was in poor condition and create a roadway where none had existed before. State and local support for the highway project was essential; federal funds were available for road improvements only as a match for state money. Local support consisted of both fund-raising and labor. Private citizens and automobile associations often had to assist official road-building crews if the work was to be finished in a reasonable period of time.

Public support for the proposed highway was tremendous. Carl Fisher remarked that the Dixie Highway was easier to promote than the Lincoln Highway because local boosters did so much of the work. Perhaps it was also because the Lincoln Highway plan seemed to be conceived and executed entirely by people living east of the Mississippi, while the Dixie Highway appealed to people living all along the route. The *Indianapolis News* printed an impassioned plea for the highway, saying that the people of Dixie were enthusiastic about a link between north and south: "To them it looms as a voice calling out of the wilderness. . . . Thousands of motorists, weary of the winter's chill, will go buzzing over the new road to seek the sunshine of the South. And when the sun beats down on the land of cotton with relentless fury, the southerners will be able to turn their motors northward where the cool breezes are."[2]

Financial supporters for the Dixie Highway volunteered funds after W. S. Gilbreath published his proposal for the route in November 1914. At first he referred to it as the Cotton Belt Route, for lack of a better name. Gilbreath urged the public to see the United States first, since the war in Europe would prevent travel abroad for years to come. Approximately one dozen men from different states each pledged one thousand dollars per year for building the highway. These "founders" enlisted the support of governors in the states through which the proposed highway would pass. Governors of interested states and the founders of the Dixie Highway movement held a preliminary meeting in January 1915 to form an association to build the road and plot a tentative route, but most of the real work was left until the next meeting.

Dixie Highway promoters had their first official meeting at the Good Roads Conference in Chattanooga, Tennessee, on April 3, 1915. More than three thousand delegates attended the meeting, which was hosted by the Chattanooga Automobile Club. Gov. Samuel Ralston of Indiana, a friend of Fisher and Gilbreath, had invited six of his fellow governors whose states lay along the proposed route. Governors from Tennessee, Kentucky, Georgia, and Indiana attended in person, Florida and Ohio sent spokesmen, and Michigan had not yet been invited to join the delegates.

Early in the meeting the governors decided to remove control of the project from the founders and give it to a board of directors, which they would appoint. Some of the founders were unhappy about this and threatened to withdraw financial support from the project, saying that they would not use their funds to assist particular people or political parties. Governor Ralston and Governor Slaton of Georgia advocated a compromise, with each of the seven governors selecting two directors, and seven other "incorporators" coming from the ranks of the original founders. Ralston chose two of his close friends as directors, Carl Fisher of Indianapolis and Sen. Thomas Taggart of French Lick. The board of directors (excluding the incorporators) was given the right to designate the route of the Dixie Highway, with advice from three highway engineers from the U.S. Office of Public Roads.

Lincoln Highway official A. R. Pardington advised the directors that they must avoid politics at all costs. He told them, "just build this road from Chicago to Miami so direct that when it is seen stamped upon a map its logic will be so strong that it will need no explanation." Before the meeting adjourned, the directors had decided to table the matter of the exact route until May.[3]

To avoid any semblance of political maneuvering, the newly formed Dixie Highway Road Commission called for each locality that wanted the route to file a brief. Commissioners asked for the following information: mileage in each county along the proposed route; condition of the road and type of surface; number and character of bridges; number of unbridged streams; plans for building and maintaining these roads; date the road could be ready for dedication; and demographic data and information on nearby historic sites. Scenic and historic places were important to Fisher, who saw the potential for tourism. Among the tourist attractions along the route of the Dixie Highway were Mammoth Cave, Abraham Lincoln's home, the Civil War battlefields of Chattanooga and Chickamauga, the old city of St. Augustine, and of course Miami Beach. On May 20, the commissioners met at Chattanooga to determine the final route. Fisher wanted the route finalized quickly, probably because his Miami Beach property could not be properly developed without a good road to bring in investors. Fisher's wife, Jane, insisted that her husband had no ulterior motive for promoting a road to Florida, and that he did not begin to develop Miami Beach as a tourist resort until 1919. Most other sources believe that Fisher had already planned to develop the area but needed the road to make his idea work.

Clark Howell, the chairman of the route committee, asked each county that wanted a part of the Dixie Highway to begin building the actual roadway (fig. 1.2).

The committee would then decide the best road segments and determine the route accordingly. So many important cities wanted a part of the highway that Fisher suggested a compromise: there would be two Dixie Highways, running parallel to each other, called the Eastern and Western divisions. The commissioners granted Indiana part of the Western Division.

In many of the southern states through which it passed, the Dixie Highway was the only good road that connected towns. It forced the construction of roads through mountain passes in Kentucky and Tennessee and spawned an interconnecting network of new roads in the southern states. But in the Midwest, geography and economics had not hindered road construction as they had in the South. The mountains of Kentucky and Tennessee had curtailed many road-building efforts, while Florida's vast swamps presented a far different engineering challenge. In contrast, Indiana had a well-developed network of roads already in place, even if the roadways were unpaved and in poor condition.

A major reason for the existence of this road network was the inauguration of Rural Free Delivery in 1896. The government would deliver mail to outlying areas, saving rural residents the

FIG. 1.2. Map of the Dixie Highway. From the collection of Martha Carver.

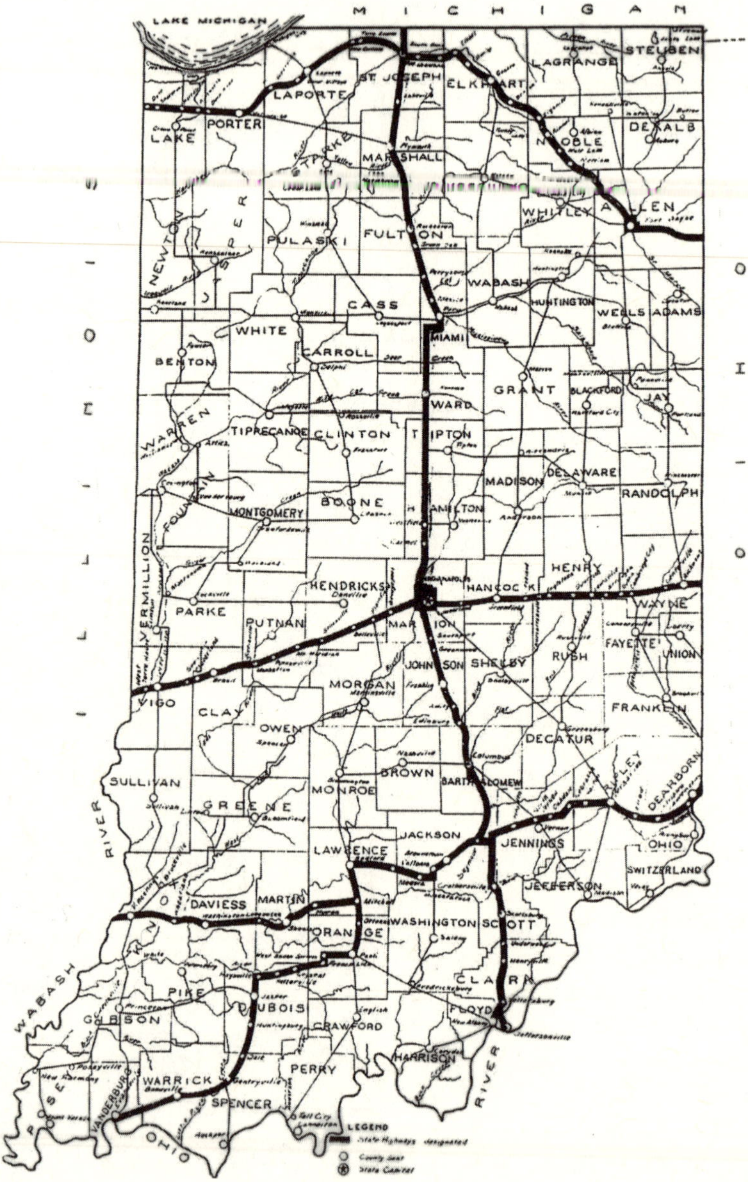

FIG. 1.3. Indiana State Highway System in 1919, as originally planned in 1917. Indiana State Highway Commission, *Portion of State Highway System* (1919).

trouble of going to the post office in town. However, mail would only be delivered to houses located on gravel or macadam roads. States were slow to develop their country roads, putting most of their effort into urban streets. In 1912 the U.S. government passed a post office appropriation bill that required the federal government to furnish one-third of the cost for road improvements; the state and local governments added the other two-thirds. Indiana formed its own Good Roads Association in 1910, which advocated establishment of a state highway department as early as 1914. The fifty thousand autos in the state were causing most of the damage to existing roads; therefore legislators thought auto owners should pay for road repair and construction. In 1913 Indiana tried to improve its roads further by adding the proceeds from automobile licenses to the highway fund. However, little was done in Indiana until mandated by the Federal Aid Road Act of 1916.[4]

By 1916 the need for improved roads was so great that the federal government passed the Federal Aid Road Act of 1916. The act provided for 50 percent matching funds for the construction of new rural public roads. Although the federal government furnished half of the cost, the individual states were to be responsible for maintaining these roads. To ensure proper distribution and use of these funds, the Federal Aid Road Act demanded that each state organize a state highway department before the state would be eligible for funding.

Indiana designed a state highway network and established a highway commission shortly before World War I, with Thomas Taggart as chairman (fig. 1.3). The Hamilton County courts objected to the commission's constitutionality in 1918, and its implementation was held up in court until 1919. The actual reason for the delay was that Hamilton County commissioners were angry that the Range Line Road (U.S. 31) was chosen for the first state highway instead of the Noblesville Road, which ran through their county seat. When the commission was reestablished in 1919, the 1917 state highway system was adopted, consisting of four major roads: roughly, State Route 1 (U.S. 31), the Lincoln Highway (U.S. 30), the National Road (U.S. 40), and U.S. 50.[5]

State Route 1, later to become U.S. 31, was an important north-south road that ran straight through the state. It connected the major industrial cities of South Bend, Kokomo, Indianapolis, and Louisville. Interestingly the first three of these were significant automobile manufacturing cities. If the purpose of the Dixie Highway was to connect the northern and southern states by the most efficient path, State Route 1 seemed to be the logical choice. At the first governors' meeting in January 1915, the suggested route did follow State Route 1 through Indiana (with a detour through Logansport), but at a later meeting delegates Taggart and Fisher voted against that route in favor of one that led in a big southwest-southeast curve.

FIG. 1.4. Thomas Taggart's hotel at French Lick Springs. Taggart's desire to attract automobile tourists to the hotel may have been one of the factors in choosing the route of the Dixie Highway. From the author's collection.

When the directors finally determined the route of the Dixie Highway, it followed State Route 1 for a short distance. From South Bend to Indianapolis, it traced the route of old Michigan Road, which, at one hundred feet, had the widest right-of-way of any road in the state. At Rochester it meandered southwest through Logansport, then to Indianapolis. On its way south it took another southwesterly turn and followed the Bloomington Road, then continued through the limestone quarries of Monroe and Lawrence counties to Paoli. There a spur ran to the resorts of French Lick and West Baden. At Paoli the Dixie Highway took a sharp bend and continued southeast to New Albany and Louisville along the old Paoli Pike, the first road in the state to be hard-surfaced. On a map the final route of the Dixie is only about twenty-three miles farther than following State Route 1, but with a Ford that averaged fifteen to twenty miles per hour, that was a substantial difference. The route passed through areas of the state that were known for their isolation and resistance to progress.

The obvious advantage of State Route 1 during the early years of auto travel was that it was a fairly straight route south from South Bend to Louisville, and, most important, it was paved for its entire length, save for a short stretch north of Perrysburg. The Dixie, on the other hand, was mostly gravel, with only a few paved sections. The portion of the Dixie between Rochester and Logansport was not even a part of the state highway system until after 1926. Why then, was this route chosen for the Dixie Highway?

One factor was Fisher's insistence on tourist attractions along the route. State Route 1 passed through an area not known for its scenic wonders or points of interest. The Dixie, on the other hand, followed the historic Michigan Road, then led to Martinsville, where middle-class tourists could visit a mineral spa or stay at the nationally known Martinsville sanitarium. Further down the road was the university town of Bloomington. The impressive quarries along the Dixie at Bedford provided limestone for many famous buildings, such as the New York Public Library and the Empire State Building. However, the main reason for Taggart and Fisher's choice of route was probably more personal: the new course of the Dixie led past Taggart's famous mineral-spring resort, the French Lick Springs Hotel, which was located about ten miles west of Paoli (fig. 1.4).

The Dixie Highway Commission had held one of its early meetings at the French Lick Springs Hotel, with Taggart as host. Taggart had been the state and national Democratic Party boss, served three terms as mayor of Indianapolis, and later served as state senator, so his hotel was a logical meeting place for Democratic politicians. His relationship with Governor Ralston was of long standing; Ralston was widely known as a member of the Taggart political machine. According to one of Taggart's employees, the governor's office had been accepting graft from Taggart and a rival hotel owner, Ed Ballard, for decades.[6] It would not be surprising if the two had exercised some political clout to get a segment of the Dixie Highway routed near Taggart's hotel.

As if the illogical choice of north-south route through Indiana were not bad enough, other factors combined to confuse the traveler as to what was or was not the Dixie Highway. As originally mapped in 1915, the Dixie had sections that ran parallel and perpendicular to the main route in every state through which it passed. It contained 3,989 miles of roads in its Eastern and Western divisions, and it ran through eight states. A few years later, the mileage had risen to 5,786. A branch of the Dixie Highway known as the Carolina Division led from Knoxville to Savannah. Georgia added a route that passed through some of its Civil War battlefields. Michigan's branch ran all around the Lower Peninsula. According to the Dixie Highway Association, all of these sections were officially considered the Dixie Highway. Beginning in 1924, the federal government solved some of the confusion by numbering all of the nation's highways according to a single system; east-west roads would have even numbers and north-south roads would have odd numbers. In response to pleas from readers, *Dixie Highway* magazine printed the state highway numbers of all the sections of its route in the November 1924 issue. In its Western Division alone, which included Indiana, the Dixie Highway incorporated sixty-six different numbered highways along the route between Michigan and Florida.[7]

During the heyday of the Dixie Highway (1914–1926), other named highway projects sought to jump on its bandwagon. This flurry of activity resulted in the

Hoosier-Dixie Highway, Dixie Route "A," the Dixie Bee-Line, and various other roads with "Dixie" in their names. Aviators could even follow the Dixie Air Line from Chicago to Miami. Although the original plan called for the Dixie Highway to run through the midwestern states from Chicago to Miami, competition for the coveted route was intense. Rather than disappoint local politicians, the Dixie Highway Association permitted each state near the proposed route to have its own section of "Dixie Highway" or similarly named route that connected its Eastern and Western divisions. The name "Dixie Highway" could signify any of these roads to the motorist not conversant with the latest "Auto Trails Map."[8]

Many of the individual roads that the Dixie Highway comprised had older or other common-use names that were more familiar to local residents. Some still continue to be used today. In Indiana the Dixie Highway included portions of several of the oldest roads in the state: Michigan Road, Crawfordsville Road, National Road, and Paoli Pike. The craze for naming routes as "auto trails" caused even more confusion. In 1923 there were more than forty named routes in Indiana. Parts of the Dixie Highway and State Route 1 bore such alternate names as Jackson Highway, Range Line, Liberty Way, National Old Trails Route, Adeway, and Ben Hur Route. When these roads had acquired still other names within a town, and all of the names were posted on street-corner signposts, the unfortunate motorist could easily become confused, make a wrong turn, and not discover his error until the next signpost.[9] It must have been difficult for motorists when they hit downtown Indianapolis, where all of these trails converged.

The official literature of the Dixie Highway Association itself added to confusion over the route. When asked to recommend the best route south, it often suggested alternate roads. *Dixie Highway* magazine printed a monthly column titled "Touring Queries" that advised the reader about current road conditions, detours, and sights to see along the way. During the first two years of the magazine's publication, the editors received queries from motorists in seventeen states. Although the magazine was primarily for motorists on the Dixie Highway, readers frequently asked the editors to recommend roads in other locations. One motorist asked, "Will you please tell me if there is an auto road between Chattanooga and Corinth, Mississippi? I know the road from Chattanooga via Nashville to Memphis, and it is no good unless you travel it in a flying machine."[10]

Parts of the Dixie Highway in Indiana also were no good without a flying machine. In 1917 the same editors recommended that motorists follow the Dixie from South Bend to Indianapolis, but then take State Route 1 from Indianapolis to Louisville because of construction and repairs to the stretches of road between Bloomington and Paoli, and New Albany and French Lick. This former stretch of road was either closed or under construction for most of the 1910s and 1920s, and even maps of the 1930s show sections that were under repair. Motorists who had no special reason to visit towns along the Dixie's route would have been foolish not to take State Route 1 instead.

Which road would one take to go to Florida for a winter vacation? Most motorists of the time probably had limited time to spend en route and wanted the best road surface and fewest construction detours. They likely wanted to stay on one road rather than try to follow the signposts pointing to a marked trail that consisted of many separate road sections (fig. 1.5). Another consideration would have been facilities for eating, sleeping, and attending to the car's needs. The Hoosier State Automobile Association published lists of approved tourist camps, more of which appeared along State Route 1 than the Dixie Highway (although unapproved camps existed along both routes).

Why were such facilities so important? In the early days of auto travel, between 1890 and 1915, facilities for motorists were scarce. Long-distance travelers carried extra gasoline in cans strapped to their cars, along with several spare tires and tubes, water for the radiator, a tool kit, and perhaps an extra battery. Motorists could not be certain of finding essential supplies, especially in rural areas; therefore they had to plan for every contingency. Gasoline stations as specialty businesses did not exist before 1907, but between 1907 and 1914 a few major oil companies began to build facilities designed especially for refueling automobiles.[11]

Early motorists did not have many choices of dining facilities before the end of World War I. Although soda fountains, cafes, and snack stands had been popular since the turn of the century, they were located only in towns and at parks or fairgrounds, not on the open road. As with other types of facilities for motorists, roadside eating places were not prevalent until after World War II. Chester H. Liebs attributes the postwar boom in roadside restaurants to people's dual desires to eat out and to ride in the car. Motorists wanted restaurants where the entire family could eat inexpensively while on a long vacation or just a Sunday drive. Tea rooms, family restaurant chains, food stands, and diners sprang up along the highways to accommodate the wave of automobile travelers.[12]

However, the motorist had to find these facilities. To promote travel, organizations published lists of facilities. The Hoosier State Automobile Association published lists of approved tourist camps, more of which appeared along State Route 1 than on the Dixie Highway, although unapproved camps existed along both routes.

Gas stations and eating places, particularly those that appealed to middle-class families, seemed to spring up more often in larger towns than in completely rural areas. Although more small towns existed along the original route of the Dixie than along State Route 1, more towns with populations of five hundred or more were sited on State Route 1, probably giving rise to more facilities for motorists. *Hoosier Motorist* magazine also listed more facilities along State Route 1. Whether there were in fact more is a moot point; what matters is that the public saw these lists

FIG. 1.5. Symbols for the marked trails in Region 2 (Indiana, Ohio, and Michigan) from the 1922 Rand McNally *Official Auto Trails Map*. These symbols appeared in Hoosier Motorist 6 in September 1918.

and just assumed that State Route 1 had better gas stations, eating establishments, and overnight accommodations.[13]

Perhaps the Dixie would have developed more efficiently had there not been a better alternative already in place. Because State Route 1 was such an excellent road compared to others in the state, there was no real external pressure for local governments to complete the Dixie Highway. Anyone who needed to get from north to south could just use State Route 1. Aside from visitors to the Springs hotels, who frequently arrived by train, few people had reason to visit the remote rural areas in the southern counties through which the Dixie passed.

Over the years State Route 1 was widened, improved, and routed around towns. There are plans for portions of its Indianapolis–to–South Bend section to become limited access. In comparison the Dixie remains two lanes in many places and is particularly unimproved between Bedford and New Albany. Of course, the northernmost portions of the two routes are the same.

The Dixie Highway consisted of hundreds of overlapping named and numbered road sections.[14] It is not surprising that motorists have varying opinions as to its exact route. From its inception in 1914 until the official federal renumbering of state highways in 1924, the Dixie Highway was one of the most conspicuous road projects in the country. Construction began in 1915 and continued throughout the 1920s.

The Dixie Highway in Indiana never seemed to develop the preeminence that it did in other states, particularly Florida, Kentucky, and Tennessee. Travelers going from Michigan to Florida, for example, frequently used State Route 1 while in Indiana in preference to the Dixie Highway until the completion of the interstate system in the 1960s. Over time many people began to think of State Route 1 as the Dixie Highway because they believed it had more of the characteristics of a major north-south connecting road: more facilities, better road surface, and a more direct route.

A sign north of Lakeville reads "DIXIE WAY ANTIQUES," which is indeed the case at that northern town. However, the Galveston quadrangle map indicates a portion of U.S. 31 north of Kokomo, Indiana, as "Dixie Highway," although only the northernmost section (South Bend to Rochester) was ever officially designated as such.[15] So in the mind of the average traveler, the Dixie Highway was the best road to the South, whatever the original planners had decided.

Even at the beginning of the Dixie Highway project, one reporter noted as fact, "regardless of the designation of certain roads as part of the Dixie Highway, travel [sic] will seek the shortest and most convenient route, road conditions being equal."[16] The current perception of the Dixie's route certainly bears out this statement. Although Carl Fisher's original Dixie Highway route itself was perhaps a failure in his own home state, the concept of the Dixie was so pervasive that its name lives on—albeit in the wrong place.

NOTES

1. Phil T. Colgrove, "The Story of the Dixie Highway," *Dixie Highway,* March 1925, 2; Jane Fisher, *Fabulous Hoosier: A Story of American Achievement* (New York: Robert M. McBride and Company, 1947), 92; Howard Lawrence Preston, *Dirt Roads to Dixie: Accessibility and Modernization in the South, 1885–1935* (Knoxville: U of Tennessee P, 1991), 97.
2. *Indianapolis News,* March 27, 1915, 15.
3. Preston, *Dirt Roads to Dixie,* 55–56.

4. James J. Flink, *The Automobile Age* (Cambridge: MIT P, 1988), 135–37, 154–56, 171. Albert C. Rose, *Historic American Highways* (Washington: American Association of State Highway Officials, 1953), 98–99, 103–4, 108, 111, 119–20; Indiana Executive Department, Division of Accounting and Statistics, *Year Book for the State of Indiana* (Fort Wayne: State of Indiana, 1917), 42; Clifton J. Phillips, *Indiana in Transition: The Emergence of an Industrial Commonwealth, 1880–1920* (Indianapolis: Indiana Historical Bureau and Indiana Historical Society, 1968), 266; Preston, *Dirt Roads to Dixie,* 19–20, has a good explanation of R.F.D. Dixie Highway directors Carl Fisher and Indiana senator Thomas Taggart were both vice presidents of the IGRA in 1912.
5. Clifton J. Phillips, *Indiana in Transition,* 266; Indiana State Highway Commission, *Map of the Indiana State Highway System* (Indianapolis: Indiana State Highway Commission, 1917); M. E. Noblet, "What Is the Matter with Indiana?" *Hoosier Motorist* 6 (April 1918): 9–12; "State Highway System of Indiana" (undated typescript in box 2, folder 9, under "Roads"), Indiana State Library; "Brief History of the Highway Commission" (typescript 1949, in box 2, folder 9, under "Roads"), Indiana State Library; *Rand McNally Official 1919 Auto Trails Map (District 2)* (Chicago: Rand McNally, 1919).
6. Douglas Wissing, "Twist the Tiger's Tail: Casino Gambling in the Springs Valley and the State of Indiana," *Traces* 10 (Fall 1998): 14–23.
7. *Dixie Highway,* November 1924, 21; *Louisville Courier-Journal,* April 30, 1916 ("Clippings"); Preston, *Dirt Road to Dixie,* 62–63; National Highways Association, *Map of the Dixie Highway* (Washington: National Highways Association, 1915). This map is located in the Map and Geography Library at the U of Illinois at Champaign-Urbana.
8. *Rand McNally Official 1918 Auto Trails Map (District 2)* (Chicago: Rand McNally, 1918); *Rand McNally Official 1923 Auto Trails Map (District 2)* (Chicago: Rand McNally, 1923); *Rand McNally Official 1926 Auto Trails Map (District 2)* (Chicago: Rand McNally, 1926); *Dixie Highway,* September 1919, 17.
9. *Rand McNally Official 1918 Auto Trails Map; Map of Indiana* (Des Moines, IA: Kenyon Company, 1923); Jeffrey L. Durbin, review of *Dirt Roads to Dixie: Accessibility and Modernization in the South, 1885–1935,* by Howard Lawrence Preston, *SCA Journal* 12 (Spring/Summer 1993): 19–22.
10. *Dixie Highway,* November 1917, 9; July 1917, 15.
11. Glory-June Greiff, "An Overview of Transportation in Indiana: 1898–1940" (unpublished report, Indiana Division of Historic Preservation and Archaeology, Indianapolis, n.d.), 22–23; John A. Jakle, "The American Gasoline Station, 1920 to 1970," *Journal of American Culture* 1 (Fall 1978): 520–42; Richard G. Minnick, "The Silent Sentinel of the American Road," *Antique Automobile* 28 (January 1964): 18–27; *Antique Automobile* 28 (March 1964): 24–32; Preston, *Dirt Roads to Dixie,* 137.
12. Chester H. Liebs discusses early roadside restaurants in *Main Street to Miracle Mile: American Roadside Architecture* (Boston: Little, Brown, 1985), 193–210.
13. *Hoosier Motorist,* 1917–1928, includes these lists of facilities in each issue.
14. *Dixie Highway* magazine contains letters from readers who were unsure of the route's exact path; in addition, its own maps did not always make clear which routes were the

official Dixie Highway and which were just good routes. *Rand McNally Official Auto Trails Maps,* 1918 and 1923 show the overlapping roads that make up the Dixie Highway's Western Division.

15. *Louisville Post,* May 29, 1915 ("Clippings"); U.S. Geological Survey, *Galveston Quadrangle,* Indiana, 7.5 Minute Series (Washington: U.S. Department of the Interior, 1959, photo revised 1980). One might infer the relative importance of the Dixie Highway in Indiana and Kentucky from the number of newspaper articles about the highway from 1914 to 1940: according to newspaper indexes, articles in the Louisville papers outnumber those in Indianapolis papers approximately five to one.
16. "Ten Months in a Fordalow," *Hoosier Motorist* 5 (April 1917): 10.

CHAPTER 2

DRIVIN' THE DIXIE HIGHWAY IN TENNESSEE

MARTHA CARVER

Tennessee played a unique role in the development of the Dixie Highway, as it was its geographic center and, with Chattanooga home to its national headquarters, perhaps the heart of the Dixie. Both the Eastern and Western Divisions stretched through Tennessee, as did a portion of the Carolina Division, crossing a total of nineteen of its ninety-five counties. Tennessee even contained one of the rare converging points of the two divisions, in Chattanooga. Even so, these ties did not necessarily generate the money or the local interest to complete the highway. Route location disputes, lack of money, mountainous terrain, and lack of interest from counties who viewed the road as more beneficial to through travelers than local residents delayed construction. The history of the Dixie Highway in Tennessee would not be complete without special emphasis on two of the most difficult sections to erect, one near Newport on the Carolina Division and the route between Chattanooga and Monteagle on the Western Division.

In the nineteenth century, roads in the rural United States usually consisted of narrow paths, often impassable in inclement weather. They contained no directions or signage, and maps were generally unavailable, so the roads were an undependable means of travel for even short distances. At that time no one seemed too concerned. People used the railroads to transport themselves and their goods for long distances. However, that complacency would dissolve with the widespread availability of the automobile. Counties struggled in the nineteenth century to build and maintain local pikes spreading outward from the county seat into the rural countryside, much like spokes on a wheel radiating outward from a hub. In contrast, the interstate highway movement of the early twentieth century emphasized interconnecting routes, not simply across county lines but across state lines, connecting the borders of the United States.

By the mid-1920s, twelve widely known interstate highways entered the South. However, the Bankhead Highway, through the Atlantic coastal states, and the Dixie Highway, along a more midwestern route, were the two most traveled.[1] Stretching from Sault Sainte Marie, Michigan, on the Canadian border to the southern tip of Florida at Miami Beach, the Dixie Highway provided a thoroughfare for the "snow birds" on their way to and from Florida. As a result of these highways, for the first time, the South became easily accessible to large numbers of people, an event that would forever alter its economy, culture, and landscape.

A substantial portion of the early usage of the interstate routes related to tourism, and a symbiotic relationship existed between the Good Roads Movement and the tourism industry. Not only did each need the other to thrive, but also at a fundamental level, each needed the other simply to survive. Therefore, it is not surprising that the routes became corridors of material culture related to tourism, "roadside architecture," and road history. Businesses that catered to the tourist or highway user abounded along the route. Small country stores flourished and provided food, supplies, and gasoline. These general businesses later faced competition from more-specialized enterprises such as service stations or diners. Tourist camps, tourist homes, hotels, and motels provided overnight accommodations. Tourist attractions such as caves, spectacular scenery and natural topographic features, Civil War battlefields, or simply warm weather and sunny beaches lured motorists south along the Dixie Highway (fig. 2.1).

FORMATION OF THE DIXIE HIGHWAY ASSOCIATION

Formed in 1915, the Dixie Highway Association quickly became one of the most influential interstate organizations in the South. Throughout its existence the Dixie Highway Association maintained its national headquarters in Chattanooga and worked in tandem with the Chattanooga Automobile Club.

Much of the early impetus for a north-south route is credited to Indiana native Carl Fisher, who is often called "the Father of the Dixie Highway." Fisher promoted a north-south interstate route to provide easy access to his real estate holdings in

FIG. 2.1. Tourist attractions frequently advertised along the roadside. This Rock City Barn was located on the Dixie Highway in Rhea County. Photo by the author.

Miami Beach, Florida, which had been little more than an isolated swamp on the southeast tip of the state when Fisher purchased it. Recognizing that a unified road system through the Midwest and the South was essential to the promotion of Florida and the success of his development, Fisher, along with Good Roads enthusiast William Gilbreath of Indiana, in 1914 began actively to promote such a highway, called the "Cotton Belt Route."

Supporters held an organizational meeting called the Governors Convention in Chattanooga on April 3, 1915, for the purpose of forming an association to promote the new road, by then officially termed the Dixie Highway.[2] More than five thousand people attended this meeting and adopted a resolution creating an organization "for the purpose of constructing a permanent highway from a point on the Lincoln Highway, near Chicago, Ill. via Chattanooga, Tenn., to Miami, Fla."[3] In May 1917 the Dixie Highway Association provisionally extended the northern terminus to Sault Sainte Marie on the Canadian border.[4] The directors officially approved this terminus in September 1921.[5]

The group elected C. E. James of Chattanooga as the first president; however, he resigned three weeks later over a routing dispute.[6] The directors then elected Judge M. M. Allison of Chattanooga as president, a position he held until the association ceased active operation in 1927. Allison was a lawyer and judge from Chattanooga who tirelessly participated in local activities and was known for his civic and social conscience. He was an energetic, forceful man who was strikingly handsome with great social charm.[7] His personality was important because Allison seems to have been "the glue" that held the association together and who drove and persuaded people to achieve the ends of the Dixie Highway Association.

The Dixie Highway would eventually incorporate multiple and indirect routes, largely reflecting the views of Carl Fisher. Since a geographically direct route would have bypassed many communities, Fisher promoted an indirect route that would benefit local towns and businesses in order to garner support for the project. It was these political considerations that led to the construction of two parallel routes with several connections, a configuration the historian Drake Hokanson described as a "wandering peavine" (fig. 2.2).[8]

Tennessee played a pivotal role geographically in the Dixie Highway system. Both branches of the Dixie Highway, the Eastern Division and the Western Division, as well as its third major section, the Carolina Division, ran through Tennessee. In addition, the Eastern and Western divisions merged to form a common alignment in only two places, Chattanooga, Tennessee, and Atlanta, Georgia. Knoxville was a key link on the Eastern Division as well as one terminus of the Carolina Division.

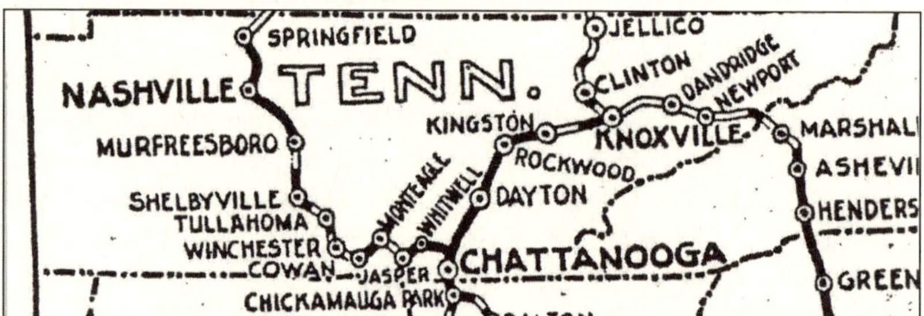

FIG. 2.2. 1923 map of the Dixie Highway in Tennessee. From the author's collection.

ROUTE OF THE DIXIE HIGHWAY IN TENNESSEE

Tennessee placed great emphasis on completion of the Dixie Highway. When the newly formed Tennessee State Highway Commission met in April 1917 to delineate the state's federal-aid roads, the Dixie Highway Association was able to convince the state to designate the Dixie Highway corridor as the state's number two road priority.[9] According to the minutes of the Chattanooga Automobile Club, "The State Highway Engineer had been instructed to . . . push the work on the Dixie Highway, giving it preference on account of its military importance."[10]

Another indication of the Dixie Highway's influence was that two communities in Tennessee named themselves after the roadway. West of Knoxville, the Eastern Division of the Dixie Highway merged with the Lee Highway, another interstate

TENNESSEE HIGHWAY PATROL BUILDING, ROCKWOOD

The 1936 Tennessee Highway Patrol building, located in Rockwood, Roane County, is an excellent example of Tennessee's response to vehicular traffic and concern for public safety along what was then its new and expanded roadway system. The building also speaks to the influence of regional highway associations and tourist traffic patterns along the roads.

During the 1920s, Tennessee greatly extended its roadway system from about four hundred miles of paved roads at the start to more than six thousand miles of state highways by the end of the decade. Regional and local highway associations acted as a strong force behind highway improvement and construction. They often organized to encourage tourism and economic incentives in their cities or states. The Dixie Highway Association, for instance, was largely responsible for the designation and construction of portions of the Dixie Highway between 1915 and 1927. This roadway traveled from Michigan to Florida, crossing through the states of Illinois, Indiana, Ohio, Kentucky, Tennessee, North Carolina, South Carolina, and Georgia.

Expansion and improvement of the roadway system led to an increase in vehicular traffic as well as to safety concerns for the mobile public along the highway. Police officers stationed in towns and cities were unable to patrol the increasing network of the state's highway system. On December 14, 1929, Gov. Henry Horton signed legislation creating the Tennessee Highway Patrol (THP). This body helped bring order and safety to the network of roads that could no longer be easily monitored by local police.

In Rockwood the street known as Kingston Avenue was designated as the Dixie Highway. This designation made Rockwood significant in tourism. The portion of Kingston Avenue known as the Dixie Highway served as the primary north-south highway through the town as well as the primary road connecting Kingston to Knoxville and Chattanooga in Tennessee, thereby requiring highway patrol facilities and officers stationed along the route. The highway patrol building in Rockwood was placed prominently on this major state route, representing not only the expansion of the roadway system, but also the state's increasing role in transportation.

The Tennessee Highway Patrol Building in Rockwood is a small one-story stone building with a stone foundation, an irregular form, a tile pyramid roof, and an interior chimney. Architecturally, the building is a fine representation of the use of Crab Orchard stone, a variety of sandstone that is quarried exclusively on the Cumberland Plateau. The Cumberland Stone Company, located in Crab Orchard, donated the stone. The Craftsman influence of the building is seen in the cut stone, use of clay tile roofing, and local materials.

The Highway Patrol Building in Rockwood served not only as a workplace, but also as living quarters for unmarried patrol officers. The back room of the building was fitted with beds and contained a kitchen and bathroom. Leon Miller, a former patrol officer who was stationed in this building for a short time, recalls that this arrangement kept

TENNESSEE HIGHWAY PATROL BUILDING, ROCKWOOD (CONT'D)

officers working day and night. He commented that the main responsibilities of the patrol officers were to enforce traffic and liquor laws. The front room functioned as an office where patrolmen did paperwork and provided tourist information.

The building served as the Tennessee Highway Patrol substation until 1952 when a new highway patrol building was constructed along a newly built highway. The construction of this new highway terminated the function of Kingston Avenue as a prominent roadway. In June 2005 the building was dedicated as the Tennessee Highway Patrol Station Museum.

There were at least three other small highway patrol buildings in the state of Tennessee. One of these buildings still exists and is located in Crossville, next to the courthouse. Another is a small brick building built in 1946 on U.S. 64 in rural Franklin County. A third building, once located in Athens, has been demolished. The Highway Patrol Building in Rockwood, one of only three known extant buildings of this type, represents an important chapter in transportation history.

AMANDA REGIER

highway, to form a common alignment into Knoxville. At this wye, a community named Dixie-Lee Junction developed. Near South Pittsburg, the Western Division of the Dixie Highway merged with the Lee Highway to form a common alignment into Chattanooga. At that wye, a community named Dixie-Lee developed, more commonly known as Kimball.

The Eastern Division of the Dixie Highway in Tennessee ran north from Chattanooga along State Route 4 (present-day State Route 29, U.S. 27) through Dayton, Evensville, Spring City, and Rockwood. In 1916 Harriman successfully petitioned to be added to the route. After leaving Harriman, the road entered Knoxville on what would later become State Route 1 (U.S. 70), a portion of which overlapped the Lee Highway.

From Knoxville the Dixie Highway passed through Maynardsville, Tazewell, and Cumberland Gap along State Highway 22 (present-day State Route 33) into Corbin, Kentucky, via Middlesboro and Pineville on present day U.S. 25E (State Route 32). However, this section was not completed expeditiously, and the Dixie Highway Association revoked the designation in May 1918. The association then designated a route from Knoxville to Corbin, Kentucky, through Clinton, Coal

FIG. 2.3. Circa 1930, brothers Henry and Elmer Nickle built this service station in the form of an airplane on the Dixie Highway north of Knoxville. Photo by the author.

Creek (later renamed Lake City), Jacksboro, Lafollette, Jellico, all in Tennessee, and Williamsburg, Kentucky, on present day U.S. 25W (State Route 9).

The Western Division ran north from Chattanooga on the Suck Creek Road (present-day State Route 27), through Whitwell, and then along present-day State Route 28 to Jasper and Monteagle. An official alternate route from Chattanooga to Jasper ran along Wauhatchie Pike, (present-day U.S. 41, State Route 2). From Monteagle the route went through Cowan, Winchester, Tullahoma, and Shelbyville on present-day U.S. 41A (State Route 16). It then followed present-day State Route 10 (U.S. 231) into Murfreesboro, and there generally followed the Old Nashville Highway into LaVergne. From there it followed present-day U.S. 41 through Nashville to Springfield and then continued north on U.S. 441 (State Route 65) into Russellville, Kentucky.

The Carolina Division began in Knoxville and passed through Dandridge and Newport before entering North Carolina. This route seems to have followed State Route 3 (present-day U.S. 25, U.S. 70, State Route 9).

CONSTRUCTION OF THE DIXIE HIGHWAY IN TENNESSEE

Each state along the route had difficulties in building the Dixie Highway, such as competing road interests, lack of financial support, swamps, and river crossings. Although road construction was always expensive and difficult, some sections were more challenging to construct than were others. Tennessee's most difficult sections were located in mountainous terrain, which was often the most scenic path through the state.

Three examples illustrate the difficulties of building the Dixie Highway in Tennessee: Knoxville, Tennessee, to Corbin, Kentucky, on the Eastern Division; Newport, Tennessee, to the North Carolina state line on the Carolina Division; and Chattanooga to Monteagle, Tennessee, on the Western Division.

KNOXVILLE TO CORBIN

The route between Knoxville and Corbin, Kentucky, was one of the most contested alignments in Tennessee. The initial route followed the Wilderness Road through Cumberland Gap. In 1775 Daniel Boone cut the Wilderness Road through the only natural opening in this area of the Appalachian Mountains. By 1800 more than two hundred thousand westward-bound settlers and explorers had gone through the gap, making it "one of the most important routes of migration in our national history."[11] The road was extremely rugged, and, for a short period in the 1890s, the conditions were so bad that lone travelers refused to use the Gap road and instead used a nearby railroad tunnel. By the early 1900s the route was a favorite place for robbers to pounce on travelers, and locals called it the "Devil's Stairway."[12]

Beginning in 1901, "Good Road Trains" traveled through the South. During these tours officials built short segments of comparatively high quality roads, essentially "seed" projects known as "object lesson roads." Officials hoped that these projects would generate support at a local level for the Good Roads Movement. Community leaders from the Middlesboro, Kentucky, area were successful in having a section of the old Wilderness Road or the Cumberland Gap Road included in the object lesson road program in 1907. It was one of only eight object-lesson roads completed in 1908, when only 680 miles of paved road existed in the United States.[13]

One of the community leaders in Middlesboro was Joe F. Bosworth, who had been elected to the state legislature in 1905 and who would become known as "the Father of Good Roads in Kentucky."[14] In a move that was partially a result of Bosworth's efforts and Middlesboro's earlier road improvements, the Dixie Highway Association routed the Eastern Division alignment from Knoxville to Corbin over the old object lesson road at Cumberland Gap. This road would later become U.S. 25E (State Route 32). Although Bell County, Kentucky, fulfilled its commitments to the Dixie Highway Association and built a seventeen-mile macadam road, other portions of this alignment were not completed. Thus the Dixie Highway Association revoked its designation in May 1918.[15] The association then routed a new alignment from Knoxville to Corbin through Clinton, Lafollette, and Williamsburg along present-day U.S. 25W (State Route 9).

In November 1924 A. P. Lerbig of Middlesboro attended a meeting of the directors in Rome, Georgia, where he gave a presentation describing the improved qual-

FIG. 2.4. A "blockhouse" facility for motorists at Cumberland Gap, along the Dixie Highway. From the author's collection.

ity of the roads in the area and persuaded the association to reinstate the Cumberland Gap route on the Dixie Highway (fig. 2.4). Ultimately the association compromised by designating the Williamsburg route as the official route and the Cumberland Gap route, which lies within the Cumberland Gap National Historical Park, as an official alternate route.[16] However, a map in an official 1925 Dixie Highway Association publication, *Up and Down the Dixie Highway,* highlights the Middlesboro route as being the preferred route.[17]

In 1996 the National Park Service completed construction of a tunnel and alternative route for motorists. As part of this project, the National Park Service used fill from the tunnel construction to raise and narrow the existing roadway with the intention of restoring it to its eighteenth-century appearance, eradicating the modern Dixie Highway and state route associations.

NEWPORT TO THE NORTH CAROLINA STATE LINE

In 1917 Tennessee designated a rugged and muddy path between Newport and Asheville, North Carolina, as State Route 3 (present-day State Route 15, U.S. 25, U.S. 70). The Dixie Highway Association designated it as part of the Carolina Division in 1918. The Tennessee State Highway Department found that "the unimproved route was a steep, narrow trail, a nightmare to travel."[18] For a short period in the early to mid-1920s, the newly formed and poorly funded Tennessee State Highway Department designated a road section in each of the state's three "grand," or major, political divisions (East, Middle, and West Tennessee) as a model road to

FIG. 2.5. The French Broad River Bridge on the Dixie Highway in Cocke County. Photo by George Hornal.

generate support for the projects. East Tennessee's model road was this section of the Carolina Division of the Dixie Highway in Cocke County.[19]

In 1920 the Tennessee State Highway Department conducted a location survey for an improved facility. However, Cocke County voted down its match for construction costs twice. At one point the Dixie Highway Association threatened to bypass Cocke County and route the highway through Greene County, but Cocke County finally approved its share of the funding in 1922. Construction was further delayed because the Southern Railway Company and the state could not reach an agreement over right-of-way along a section where the terrain forced the highway to parallel the railway for more than a mile along the banks of the French Broad River. After a flood in 1920, the state's engineers relocated a five-mile stretch of the road in an effort to minimize future flood damage. Concurrently the state decided to relocate the road so that it crossed the railway once perpendicularly rather than following the railway in a parallel line. This resulted in the need for a major new bridge structure to span both the French Broad River and the Southern Railway.

In December 1922 the state let a contract for Federal Aid Project Number 23, a sixteen-foot-wide macadam road ten miles in length from near Del Rio in Cocke County to the North Carolina state line. Rather than detour a hundred miles to avoid construction activities, motorists made their own roads through the area, although they were often impassable because of mud. Many of the motorists were tourists, "most of whom were accustomed to smoother country, and better roads, and these were far from bashful in their caustic criticism." That year farmers along the way "harvested the automobile crop; one even charged motorists a dollar for crossing his land."[20] However, Cocke County did not agree to provide its match for the French Broad River bridge until 1926, and the state did not complete the open spandrel concrete arch bridge until February 1928 (fig. 2.5).

CHATTANOOGA TO MONTEAGLE

The route between Monteagle and Jasper, which crossed the Cumberland Plateau at Monteagle Mountain, and the route between Jasper and Chattanooga, which crossed Walden's Ridge, were probably the two most difficult sections in Tennessee to complete. The route climbed and descended steep mountains to cross the floor of the Sequatchie Valley, only to climb and descend the mountain across the valley floor. In addition to the terrain, political and funding problems hindered these sections. Part of the problem was that both of these road segments crossed steep mountains that were located on the edge of Marion County. Not only were the roads difficult and expensive to construct, but local voters perceived those sections as benefiting through travelers and as having little value for residents. Therefore the project received limited support in Marion County, which had to pay a portion of the cost.

Although an older road from Monteagle to Cowan provided northward access, prior to the creation of the Dixie Highway, no road existed from Monteagle south to Jasper, a route that involved an extremely steep and rugged descent from Monteagle Mountain. The Dixie Highway Association chose a route south from Monteagle to Jasper that required a new road rather than the more typical overlay approach. It is unclear why the Dixie Highway Association chose this route rather than using the existing road that ran east from Monteagle to Tracy City and then south to Whitwell. A local newspaper article in 1976 stated that the Grundy County Court refused to fund improving a road project along this route and therefore lost the Dixie Highway and, as a consequence, eventually lost U.S. 41. "This one decision by the court probably did more than any other in the county's history to determine the type of place it is today."[21] However, the Grundy County court minutes do not contain a specific reference to this. Another possibility is that since a route through Grundy County bypassed Jasper, the Marion County seat, the Dixie Highway Association may have deliberately chosen the Monteagle to Battle Creek (Tennessee) route because it came through Jasper and would have thus generated more support at a local level.

The route from Monteagle to the head of Battle Creek was a distance of four and one-half miles and covered a difference in elevation of approximately twelve hundred feet.[22] Few crossings cut through the mountain range, and this section "made or broke" travel between Chattanooga and Nashville, making it a pivotal component of the Western Division. In 1916 Marion County sold bonds and began road construction but quickly ran out of money. In 1918 the State Highway Department graded the road, but rain soon washed deep ditches through the dirt lane, and the State Highway Department had no maintenance authority to repair it. The motorists who tried crossing the rutted, washed-out path over the mountain warned their friends, and the mountain crossing became an obstacle dreaded by tourists days

FIG. 2.6. As this editorial cartoon in the March 1918 issue of the *Dixie Highway* illustrates, the Dixie Highway Association tied road construction to efforts to win World War I in an attempt to secure funding for completion of the Dixie Highway.

FIG. 2.7. View of the Dixie Highway approaching Monteagle Mountain. The text on the postcard reads, "Approaching the Cumberlands on the Dixie Highway." From the author's collection.

before they reached it. Once they crossed it, they had "no words with which to express their horror of such a road" (fig. 2.7).[23]

By the early 1920s, this section was the most widely known road problem in Tennessee.[24] Even when the state assumed primary responsibility for the road, Marion County at first declined to provide even matching funds for the road. The county finally approved a bond issue in 1923, and the state finished this section in 1925.[25] A special committee from the state senate inspected the finished road in 1925 and reported that it had been able to drive from Nashville to Chattanooga in only five and one-half hours. This event was so noteworthy that the WSM radio station in Nashville, a wide-ranging station heard throughout most of the eastern United States, began to broadcast that the road was no longer "a dangerous mule trail."[26] Today the east-bound route of Interstate 24 overlays the route of the Dixie Highway on the east side of the mountain, from Monteagle to Battle Creek, but Monteagle remains a difficult and dangerous crossing.

From Jasper near the foot of Monteagle Mountain to Chattanooga, motorists faced two possible routes: the Suck Creek Gorge route or the Wauhatchie route. Both routes had significant problems. The Suck Creek Gorge route was longer and crossed the steep and rugged Walden Ridge, but it crossed the Tennessee River in downtown Chattanooga across the substantial Market Street Bridge, a double-leaf bascule lift bridge built between 1914 and 1916. The Wauhatchie route required the motorist to cross the Tennessee River on the Rankin (Kelly) Ferry. It also required access around and over Raccoon Mountain as well as access over or around Lookout Mountain.

The Dixie Highway Association selected the Suck Creek route through Whitwell and over Walden Ridge. This section, the Suck Creek Gorge route, seems to have been a special project of the Dixie Highway Association and the Chattanooga Automobile Club. The Dixie Highway Association even managed to secure a special state enabling act that allowed Marion County to pass a bond issue to fund this section and to change the Hamilton-Marion county line to allow Hamilton County to provide additional funding.

Not surprisingly, no road over this section of Walden Ridge existed until the early twentieth century, when Marion County built a road on the west side to Ketner's Gap. However, the county did not have enough money to extend the road to the east toward Chattanooga through the extremely rough terrain of the Suck Creek Gorge. Marion County appropriated twenty-five thousand dollars for the Suck Creek Gorge section in a joint project with Hamilton County that used 150 state convicts to build a narrow graded road. Unfortunately the road was still in poor condition, and in 1919 the Dixie Highway Association successfully lobbied to designate the Suck Creek Road as a federal aid project in Tennessee. The state expanded the project in 1921 to include the improvement of the earlier road from Whitwell to Ketner's Gap.[27] In 1923 the state assumed responsibility for the construction of this section and completed it circa 1925.[28]

The Suck Creek Road section was considered to be the highest point on the entire Dixie Highway route and roughly its midpoint as well. These factors, as well as in recognition of the difficulty in building this section, led the association to choose it to be the location for a park and monument commemorating President Allison's contributions to the Dixie Highway (fig. 2.8).

The Suck Creek section of the Dixie Highway was—and is—a steep, winding road with hairpin curves that hug rock outcroppings. It provides a bird's-eye view of the Sequatchie Valley on the west side of the mountain and a spectacular view of Signal Mountain and the boulder-strewn Suck Creek Gorge on the east side. Even after the exhaustive work of the Dixie Highway Association on this section, its rough topography guaranteed that motorists continued to have problems traversing it.

In response to the problems with the Suck Creek route, the Dixie Highway Association designated the Wauhatchie route as an official alternate route. Between 1930 and 1932 the state designated the Wauhatchie route as U.S. 41, and the Suck Creek Road section with the Allison Monument lost its U.S. routing designation. Although the terrain was not quite as rough as that of the Suck Creek Gorge route, the alternate Wauhatchie route was also in poor condition and needed extensive improvements.

Judge Will Cummings of Hamilton County was a member of both the Dixie Highway Association and the Chattanooga Automobile Club; he lived near Wauhatchie and was an activist in the Good Roads Movement and instrumental in the con-

FIG. 2.8. The circa 1924 memorial commemorates Judge M. M. Allison, first president of the Dixie Highway Association and one of the people most responsible for its success. Photo by the author.

struction of the Market Street Bridge in Chattanooga. With his political connections Cummings was repeatedly able to secure state and federal aid projects for the area.[29] Cummings organized efforts to apply for federal moneys to build the Wauhatchie Pike, and when local bonds failed to raise Hamilton County's required portion, Cummings borrowed five thousand dollars under his own name to secure the federal aid grant.[30] At that time local newspapers cited Wauhatchie Pike (Dixie Highway), one of the first concrete roads in Hamilton County (fig. 2.9), as the tenth federally financed road in the United States and as the first federal aid project in the South.[31] After the state designated Wauhatchie Pike as U.S. 41, the state rebuilt it between Jasper and Chattanooga. The completed twenty-four-mile route, which the state legislature designated as the Will Cummings Highway, opened in 1935. Markers commemorating the highway and Cummings are located in Jasper and at two locations at the base of Lookout Mountain. President Franklin D. Roosevelt participated in elaborate dedication ceremonies on October 24, 1937, by pushing a gold-plated button from his office to transmit a signal.[32]

END OF THE DIXIE HIGHWAY ASSOCIATION

The Federal Aid Act of 1916 is generally seen as the culmination of the Good Roads Movement. Although the act did not immediately denote the end of the highway associations, federal funding and statewide highway programs did change the associations. Once states formed highway departments, the counties and associations saw them as allies, not rivals, and lobbied extensively for local routes to be included on the state's system in order to acquire state or federal funds to finish the highway. In 1922 the Dixie Highway Association decided to focus on "securing the cooperation" of each state highway department along the Dixie Highway to complete one line and then to focus on other branches.[33] In general state highway departments with federal or state funds, not individual counties or highway associations with private funds, built the actual roadway of the Dixie Highway.

FIG. 2.9. A view of the 1918 Wauhatchie Pike. The 1937 reconstruction bypassed this section at Jonas Bluff, leaving it as a small pull-off with the original concrete paving. From the author's collection.

By 1926 about six hundred highway associations existed in the United States with roughly 70 percent of their routes overlapping.[34] In addition the various state highway departments had designated many roads as state routes beginning in about 1915–18, but these designations usually had no continuity across state lines. Together these competing road names and numbering systems created havoc for travelers. In 1925 the American Association of State Highway Officials recommended, and the United States Highway Board adopted, a program to designate official U.S. Highways with a numbering system instead of names. This plan was completed by the fall of 1925 and remains essentially intact today.

Ironically the demise of the Good Roads Movement and highway associations is fundamentally tied to its success. In 1915, when Carl Fisher and others formed the Dixie Highway Association, road conditions were typically deplorable throughout the South. The Federal Aid Road Act of 1916 and the Federal Highway Act of 1921 pumped a total of $150 million at a 50 percent match into road construction across the country, and by the mid-1920s, road conditions across the South had greatly improved. This included not only the named highways like the Dixie but also a much larger network of crisscrossing local roads. For instance, in 1922 the state of Tennessee maintained only 244 miles of state roads, and road construction ranked fourth in state spending. However, under Gov. Austin Peay, Tennessee's "Road Building Governor," the state in 1923 spent more money on road construction than on any other single program. And, beginning in 1924, Governor Peay's administration spent more money on road construction than it spent on all the other programs combined.[35] By 1926 Tennessee's state route system contained a strong network of 6,000 miles of interconnecting, relatively high-quality roads.

The U.S. routing designation, which fragmented most named highways, essentially killed the highway associations.[36] The Dixie Highway Association was no exception. Beginning in 1925, the Dixie Highway minutes contain discussions about the organization's financial problems and concerns about continuing publication of its magazine. In July 1926 the association noted that the advertising was insufficient to pay for publication of its monthly magazine, the *Dixie Highway.* With its purpose of a route from midwestern Canada to Florida accomplished, the Dixie Highway Association ceased active operation in 1927.[37] Its longtime supporter and ally, the Chattanooga Automobile Club, essentially absorbed it. In April 1976 the Chattanooga Automobile Club merged with the Mid-South Club and officially ceased to exist although it continued to operate as the Chattanooga branch office.[38]

NOTES

1. Howard Lawrence Preston, *Dirt Roads to Dixie: Accessibility and Modernization in the South, 1885–1935* (Knoxville: U of Tennessee P, 1991), 129.
2. Dixie Highway Association, "Minutes" (1915): 1 (archived at Chattanooga Automobile Collection, Chattanooga–Hamilton County Bicentennial Library); Preston, *Dirt Roads to Dixie,* 54.
3. Dixie Highway Association, "Minutes" (1915): 1.
4. Ibid. (1917): 51.
5. Ibid. (1921): 85–86.
6. Ibid. (1915): 21. Martin Ochs, "Dixie Highway Trailblazer, Judge M. M. Allison" (speech given to the Chattanooga Area Historical Association, November 20, 1962), Chattanooga Automobile Collection, Chattanooga–Hamilton County Bicentennial Library.
7. Ochs, "Dixie Highway Trailblazer," 4.
8. Drake Hokanson, *The Lincoln Highway* (Iowa City: U of Iowa P, 1988), 21.
9. Tennessee State Highway Commission, Nashville, "Minutes" 1 (July 1915–March 1919): 97 (State Highway Department Collection, Tennessee State Library and Archives, Nashville). The state designated as its first priority the Memphis to Bristol Highway, later designated as State Route 1 and U.S. 70.
10. Chattanooga Automobile Club, "Minutes" 17 (January 1918), Chattanooga Automobile Collection, Chattanooga–Hamilton County Bicentennial Library, Chattanooga.

11. Stanley Folmsbee, Robert Corlew, and Enoch Mitchell, *Tennessee: A Short History* (Knoxville: U of Tennessee P, 1969), 241.
12. Robert Kincaid, *The Wilderness Road* (New York: Bobbs-Merrill, 1947), 27–32.
13. Ibid., 352.
14. Ibid., 352, 358.
15. Dixie Highway Association, "Minutes" (1919): 62–63, 66.
16. Ibid. (1924): 160.
17. J. Newcomb, *Up and Down the Dixie Highway* (Brooklyn, 1925).
18. Leland Johnson, *Memphis to Bristol: A Half Century of Highway Construction* (Nashville: Tennessee Road Builders Association, 1978), 33.
19. *Newport Plain Talk,* July 7, 1925, July 1 and 5, 1927.
20. Johnson, *Memphis to Bristol,* 34.
21. Gary Matthews, "Mountain Highway and Byways," *Grundy County Herald,* September 2, 1976.
22. Tennessee Department of Highways and Public Works, "More about Monteagle," *Tennessee Highways* 3 (July 1924): 13.
23. Dixie Highway Association, *Dixie Highway,* February 1924, 20. Johnson, *Memphis to Bristol,* 37.
24. State of Tennessee, *Biennial Report of the Commissioner of the Department of Highways and Public Works for the Years 1923 and 1924* (Nashville: State of Tennessee Department of Highways and Public Works, 1924), 7.
25. S. E. Moreland, "How Is the Mountain Road at Monteagle?" *Tennessee Highways* 3 (June 1924): 12.
26. Johnson, *Memphis to Bristol,* 38–39.
27. J. E. Moreland, "The Suck Creek Road," *Tennessee Highways* 2 (October 1922): 39.
28. Chattanooga Automobile Club, "Minutes" (March 29, April 5, May 24, and July 7, 1917).
29. Fred Hixson, *The Age of Will Cummings* (Chattanooga, 1962), 50–52.
30. Ibid., 50.
31. Ibid.; *Chattanooga Times,* February 25, 1918; March 10, 1969.
32. Hixson, *Age of Will Cummings,* 109.
33. Dixie Highway Association, "Minutes" (1922): 103.
34. Ibid. (1926), 196. Clarence Kolwyck, "The Chattanooga Automobile Club and the Dixie Highway Association," *Motorist* (July/August 1976): 5, 11.
35. Joseph T. Macpherson, "Democratic Progressivism in Tennessee: The Administration of Governor Austin Peay," Ph.D. dissertation, Vanderbilt U, 1969, p. 194.
36. Hokanson, *Lincoln Highway,* 108.
37. Marion County Quit Claim Deed, 1924, Deed Book 9–A:21, 1979.
38. Kolwyck, "Chattanooga Automobile Club," 5, 11.

CHAPTER 3

HEADING SOUTH WITHOUT GETTING SIDETRACKED: THE DIXIE HIGHWAY IN GEORGIA

JEFFREY L. DURBIN

The two main divisions of the Dixie Highway passed through eight states: Michigan, Illinois, Indiana, Ohio, Kentucky, Tennessee, Georgia, and Florida, with the Carolina Division passing through North Carolina, South Carolina, Tennessee, and Georgia. Having all three major routes pass through Georgia meant an opportunity for more businesses to profit, but it also resulted in an enormous financial burden on the state to complete the road. With the Dixie Highway Association's leadership fighting over the route and with disagreements among local interest groups further exacerbating the issue, the story of the Dixie Highway in Georgia perhaps epitomizes the axiom that all politics are local. Similar to the dirt roads Union troops marched on during their 1864 invasion of Georgia, the Dixie Highway brought northern developers and tourists through the state in large numbers, helping to transform not only the landscape but also Georgia's culture.

One of eight states the Dixie Highway ran through, Georgia was pivotal for automobile tourists making their way to South Florida. Clearly the highway held much promise not only for Carl Fisher, with his plans to create a tourist mecca in Miami Beach, but also for every community along its route. Georgia businessmen and town boosters soon recognized the potential benefits of the Dixie to economic vitality. Without hesitation its residents began expending much time and effort to construct the permanent road envisioned by the Dixie Highway Association in 1915.[1]

Similar to conditions in other southern states, Georgia roads were primitive by the standards of the day and road builders faced a formidable task in creating highways suitable for long-distance automobile travel.[2] In addition to the conquering the state's physical conditions, which included mountainous terrain in its northern section and broad river floodplains in its southern region, its citizens would need to overcome the inadequate financial resources of the state. The Dixie Highway covered much more territory in Georgia than in the other seven states. By 1924 the Peach State had approximately 1,300 miles of the highway's total 5,786 miles. But because of the Dixie's shifting alignments through the state, at one time or another Georgia had 1,500 miles, more than any other state had.[3]

While they could boast about how much visitors would see on trips through their state, Georgia's citizens also had much work ahead of them in creating a clearly routed highway that tourists could easily navigate. As important as having a suitable road for automobile travel was determining the route that the Dixie would take through Georgia and ensuring it was marked at regular intervals. Unfortunately tourists likely had as much difficulty finding the route as they did driving treacherous roads. Unlike the Lincoln Highway, the Dixie would not have a single alignment.

The reason that the Dixie Highway would have multiple routes stemmed from competing priorities within the highway association. In May 1915 two association board members, *Atlanta Constitution* editor Clark Howell and Carl Fisher, fought over the route. Howell wished to postpone deciding where the road would run, to encourage counties along the highway to improve the local roads, while Fisher wanted to establish an alignment for the Dixie to maintain momentum for completing the highway. As a compromise the association created two divisions for the Dixie Highway, which included different routes through Georgia. Both of Georgia's routes—the Western and Eastern divisions—began in the northwest corner of the state and continued southward before entering Florida.[4]

From where the two divisions split in Chattanooga, the Western Division of the Dixie Highway entered Georgia on present-day U.S. 27 near Rossville.[5] The highway then passed through the town of Fort Oglethorpe and the Chickamauga-Chattanooga battlefield, which became the nation's first national military park in 1890.[6] That the Western Division would be routed on the main road through the Civil War battlefield was intentional. A system of improved roads existed in the park since its establishment, and compared to the dirt roads outside the park, Chickamauga's gravel road was an excellent corridor.[7]

After leaving the Chickamauga battlefield, tourists traveling the Dixie's Western Division passed through the towns of Rock Spring, Noble, Lafayette, Trion, Summerville, Crystal Springs, Armuchee, and Rome. Apart from several bypassed sections of the Dixie, present-day U.S. 27 largely follows the route of the named highway.[8] At Rome, however, the route turned eastward where travelers followed present-day State Route 293 to Kingston and Cass Station, the junction of the Dixie's Western and Eastern divisions. After leaving Cass Station, automobile tourists on both of the two main divisions of the Dixie Highway headed south to Cartersville following the same alignment on State Route 293.

From Cartersville, travelers using both the Western and Eastern divisions drove on a single alignment, now designated as State Route 293, through Emerson, Allatoona, Acworth, Kennesaw, Kennesaw Mountain National Battlefield Park (established 1917), and the town of Marietta. Near where State Route 293 intersects with the current route of U.S. 41, the Dixie entered Marietta on Kennesaw Avenue. South of the town, the highway continued on the present-day Atlanta Road before it intersected with West Paces Ferry Road. Prior to entering the village of Vinings, tourists crossed the Chattahoochee River on a steel truss bridge.[9]

From Vinings the Dixie Highway followed West Paces Ferry Road before it intersected with Peachtree Road in Buckhead. From there the highway entered Atlanta on Peachtree Road and followed it into downtown Atlanta.[10] In Atlanta the Dixie's two divisions split again. The Western Division continued south on present-day U.S. 19/41 through Jonesboro.[11] Just south of the town, the division followed a road that lies east of present-day U.S. 41. Using the older road as they continued into Spalding County, travelers went through the towns of Sunny Side, Pomona, and Experiment before entering the county seat of Griffin.[12]

Although the present alignment of U.S. 41 bypassed it, the old route of the Dixie Highway north of Griffin is significant. In 1920, Spalding County paved a fourteen-mile-long section, becoming the first county on the Dixie Highway between Chicago and Miami to have improved the route with a continuous hard-surfaced road. This project cost seven hundred thousand dollars, half of which the county provided in order to match the federal share of the endeavor. To mark the event, Spalding County commissioners staged a motorcade over the route and held a luncheon.[13]

Leaving Griffin, automobile tourists on the Western Division drove on what is now a county road to Orchard Hill, Milner, and Barnesville. From Barnesville, the Western Division turned eastward to Forsyth on what is now U.S. 41. South of Forsyth the Dixie then ran southerly to Macon, also on present-day U.S. 41. At Macon the alignment of the Western Division again joined with the path of the Eastern Division.[14] South of the town, the two divisions again split. From there the Western Division took travelers to the middle Georgia towns of Fort Valley, Montezuma, Oglethorpe, Andersonville, and Americus on present-day State Route 49. After Americus tourists driving the Western Division traveled on what is now U.S. 19 to Albany, Camilla, and Pelham. South of the Mitchell-Thomas county line at the town of Meigs, the route of the Western Division followed present-day State Route 3 to Ochlocknee and Thomasville. From Thomasville travelers continued southward on what is now U.S. 319 and after leaving Moncrief entered Florida.[15]

Early maps indicate that the Dixie Highway's Eastern Division entered Georgia on the same alignment as the Western Division before splitting and heading eastward to Ringgold, where it turned south toward Tunnel Hill.[16] However, sometime after 1926, the route of the Eastern Division was moved to the alignment of present-day U.S. 41 and entered Georgia north of Ringgold.[17] Between Chattanooga and Atlanta the Dixie's Eastern Division parallels the route of another travel corridor—the Western and Atlantic (W&A) Railroad, taking advantage of the towns that had developed along the rail line.

Occasionally the Dixie's Eastern Division through Georgia is referred to as the "Battlefield Route" or the "Johnston-Sherman Link"[18] of the Dixie Highway. The latter name memorialized Confederate general Joseph E. Johnston and Union general William T. Sherman, whose armies waged a series of battles in North Georgia during the spring and summer of 1864 (fig. 3.1). The Eastern Division closely followed the invasion path that Sherman took toward Atlanta, while portions of the highway between Atlanta and Savannah parallel the route of his destructive "March to the Sea."[19]

From Ringgold, the Eastern Division largely follows the route of present-day U.S. 41. Among the towns of Tunnel Hill, Rocky Face, Dalton, Resaca, Calhoun, and Adairsville, the highway winds through the hilly countryside (fig. 3.2). As with the Western Division, there are a few places in North Georgia where the numbered route bypassed the Dixie's Eastern Division. For example, such segments may be

FIG. 3.1. Joseph E. Johnston monument, Dalton, Georgia. From the author's collection.

ATLANTA CAMPAIGN WAYSIDES

In January 1900, Cornelius Cadle, a former colonel with the Army of the Tennessee, wrote to T. H. Martin of the Atlanta Businessmen's League, suggesting the erection of historical markers along the main battle lines of the Atlanta campaign through northern Georgia. Though Cadle's suggestion generated some interest, no serious efforts to erect markers about the campaign were made until the 1930s, when the National Park Service (NPS) began negotiating with Georgia officials to acquire property for the erection of five historical markers.

In 1937 the Department of the Interior appropriated ten thousand dollars for the construction of five historical markers and small parks in northern Georgia, using labor and services from the NPS and the Works Progress Administration (WPA). Throughout 1938, the NPS, the WPA, and Georgia officials worked together to locate desirable spots to construct the commemorative parks, which would overlook conspicuous portions of battlefields or encampments. The areas chosen were located on Highway 41, the Dixie Highway, which was the most heavily traveled road between Chattanooga and Atlanta, and were highly visible locales. Landscape architects employed by the NPS drew up plans for the five parks, which were alternatively called "exhibits," "markers," and "waysides." Plans called for a stone wall enclosure with picnic tables and a bronze plaque installed on a flagstone to interpret the movements of the armies in the vicinity.

A variety of problems obtaining the land in Ringgold Gap, Rocky Face, Resaca, New Hope, and Cassville delayed the beginning of construction of any of the projects until late 1939, when the project at Ringgold Gap was begun under auspices of the WPA and NPS. However, New Deal programs such as the WPA were being dismantled by 1940, and little progress was made on the wayside at Ringgold Gap or at any of the other sites. During World War II, the NPS concentrated its efforts in northern Georgia on the national military parks at Chickamauga/Chattanooga and Kennesaw Mountain and provided little or no maintenance to the sites that had been selected for waysides.

After 1945 the NPS resumed planning for the Atlanta Campaign Waysides and employed caretakers to manage the four sites where construction had begun. Although the sites were maintained through 1946, no funding for additional construction was appropriated. Construction resumed in early 1947, when NPS hired the artist John Steinichen to design the bronze markers for the waysides. James Clark quarried the stone for the walls of the waysides at a Georgia rock quarry. By June the NPS could report that "the stone wall and flagstones work at the Ringgold site has been completed," marking the completion of the first of the five Atlanta Campaign Waysides.

In addition to the interpretative marker on the flagstone, the NPS and the state of Georgia each installed vertical historical markers detailing the movements of the armies during 1864. The Atlanta Campaign Waysides served as roadside parks and rest areas for travelers along the north Georgia portion of the Dixie Highway until most through-traffic was diverted from the Dixie with the construction of Interstate 75.

TARA MITCHELL MIELNIK

FIG. 3.2. U.S. 41 near Dalton, Georgia. From the author's collection.

found in Whitfield County south of Dalton, in Gordon County south of Calhoun, and in Bartow County north and south of Adairsville as well as north and south of Cartersville. The present alignment of U.S. 41 also bypasses the downtown areas of cities such as Dalton and Cartersville, and smaller towns such as Tunnel Hill and Adairsville.[20] Similarly U.S. 41 bypasses the route of the Eastern Division north of Cassville, but the numbered route originally followed what is presently a county road into the town.[21] Near Cass Station the Eastern and Western divisions joined together and followed a unified alignment to Atlanta as described above.[22]

After Atlanta automobile tourists traveling on the Eastern Division of the Dixie Highway in 1915 would have gone south through McDonough and Jackson on present-day U.S. 23 before reaching Indian Springs (fig. 3.3) and Forsyth (now State Route 42) and then to Macon. It is unclear just what road the division followed between Forsyth and Macon. Not until February 1916 did the association's board of directors choose the route of the Eastern Division beyond Macon, when it adopted a Clark Howell resolution. According to that, the Eastern Division would leave Macon before it continued northeastward to Milledgeville, Sandersville, and Louisville. From Louisville the route then took travelers to Waynesboro, Sylvania, Springfield, Savannah, Darien, Brunswick, Woodbine, and Camden before crossing into Florida just north of Yulee.[23]

Unfortunately no Dixie Highway Association map from 1916 has ever been found, and the 1917 map lacks sufficient detail to determine the precise route of the Eastern Division between Macon and Milledgeville.[24] On the lawn of the Jones

FIG. 3.3. New Hotel Elder, Indian Springs, Georgia. From the author's collection.

County Courthouse in Gray is a cross-shaped concrete marker with the inscription "Dixie Highway." If the marker was initially placed in Jones County, the original alignment of the Eastern Division was likely near the alignment of present-day U.S. 129 between Macon and Gray.[25]

From Gray the highway likely followed present-day State Route 22 to Milledgeville. From there the Eastern Division took travelers to Sandersville on what is now State Route 24. Beyond Sandersville the highway continued on State Route 24 toward Louisville and Waynesboro. Seven miles north of Sylvania, the highway went southward on present-day U.S. 301 in the town. From Sylvania the division followed along what is now State Route 21 to Springfield and Savannah.[26]

South of Savannah, travelers on the original alignment of the Dixie Highway's Eastern Division used the route of present-day U.S. 17. On this segment of road, automobile tourists drove through the towns of Midway, Darien (fig. 3.4), and Brunswick. From Brunswick cars headed southward to Woodbine and Kingsland before entering Florida at the St. Marys River. Prior to 1927, when a steel truss bridge over the river was constructed, travelers crossed the St. Marys River on a ferry.[27]

To make matters more confusing, less than four months after it determined the route of the Eastern Division south of Macon, the Dixie Highway Association again changed the road's alignment south of Atlanta. Because the counties the 1915 route passed through south of the city were doing little to maintain "a well surfaced and drained all year round road," in July 1916 the Dixie Highway Association's board of directors "abrogated" the original route of the branch.[28]

FIG. 3.4. Darien, Georgia. From the author's collection.

In its place the directors plotted a new route for the Eastern Division that took drivers east of Atlanta to Waynesboro, where the highway's new route would join with the alignment of the division that the board had approved earlier that year. Rather than head in a southerly direction, the new alignment of the Eastern Division went due east of Atlanta through the towns of Decatur, Stone Mountain, Covington, and Madison on present-day U.S. 278, but after Madison the route followed what is now U.S. 441 to Eatonton and Milledgeville. From Milledgeville the revised route followed the path that the board had adopted earlier that year, and it continued on to Sandersville, Louisville, Waynesboro, Sylvania, Springfield, Savannah, Darien, Brunswick, Woodbine, and Camden.[29]

The problems with the routing of the Eastern Division were not completely solved, as the route was shifted a third time. The Dixie Highway Association's 1924 map shows that the Eastern Division followed the same route as the Western Division as far south as Macon. From there the 1924 route of the Eastern Division essentially followed what is now U.S. 41 to Perry. South of Perry, the route took travelers to Hawkinsville on present-day U.S. 129 to Abbeville, Fitzgerald, and Ocilla. From the last town, tourists driving the 1924 alignment followed what is now State Route 32 to Douglas. After following present-day State Route 158 through the Talmo, the Eastern Division followed what is now U.S. 82. Just west of Waresboro the route took a bypassed section of U.S. 82 to Waycross. From there the alignment of the Eastern Division is synonymous with present-day U.S. 1/23.

North of Folkston, the route followed what is now U.S. 301, taking auto tourists into Florida.[30]

Adding to the tangle of the two main core routes of the Dixie Highway in Georgia, there were also other alignments. The most important of these was the Carolina Division. Although the alignment and the route's southern terminus changed while the Dixie Highway Association was active, the northern terminus of the Carolina Division remained unchanged. The Division entered Georgia on present-day U.S. 25 at Augusta. South of Augusta, the route continued on what is now U.S. 25 south of Waynesboro, where it merged with the Eastern Division route.[31]

A Central Division, running between the Western and Eastern divisions south of Macon, existed in 1916. By 1924 the route for the Central Division had become the new alignment of the Eastern Division south of Macon when the original alignment for the 1916 Eastern Division south of Waynesboro became subsumed under the Carolina Division.[32] During its May 1922 meeting in Jacksonville, Florida, the highway association adopted a fourth alignment of the Dixie that originated in Perry and continued due south through Unadilla, Vienna, Cordele, Ashburn, Tifton, Sparks, and Valdosta before it entered Florida north of Jasper. While no designation appears to have ever been given to this alignment, the route eventually became present-day U.S. 41 and parallels the modern super highway Interstate 75 through South Georgia.[33]

Aside from having multiple corridors running through the state, the routes of the different alignments through Georgia changed during the existence of the Dixie Highway Association and after the organization disbanded in 1927.[34] These shifts appear to have been the failure of counties along the routes to maintain their share of the road. For example, in 1916, the association threatened to reroute the Western Division in North Georgia when Bartow County had failed to maintain its section of the road.[35]

With so much of the highway's total mileage, the Peach State was the beneficiary of having 53 of its 161 counties at one point or another along the various routes of the Dixie Highway.[36] Georgia, more than any other state, characterized the description of the Dixie Highway as a "wandering peavine."[37]

Whatever the branch, promotion of the Dixie Highway in Georgia began soon after the formation of the highway association, and much of this promotion was generated by the association director Clark Howell and his newspaper, the *Atlanta Constitution*. Howell used the it as a mouthpiece for his highly vocal support of the Good Roads Movement in general and of the Dixie Highway specifically.[38] For example, during the month of March 1915, the newspaper had twelve articles and one editorial that generated much enthusiasm in Atlanta and Georgia, even before the fledgling association held its April 3 meeting in Chattanooga.[39]

In describing the promise the highway held for Georgia and the widespread support for improving the route in the Peach State, the daily's reporters wrote with gusto. For example, the newspaper reported in 1915 that the Dixie Highway would rival the transcontinental Lincoln Highway in its success. The paper was effusive in its comparison of the importance of the Dixie to highways in ancient Rome.[40]

During an association-sponsored motorcade running from Chicago to Miami, a front-page article in the October 15, 1915, edition of the *Atlanta Constitution* invited the city's automobile owners to turn out and give the "tourists the greatest road reception they have received anywhere along their route."[41] Other Georgia newspapers were active in promoting the Dixie Highway, especially when the routing of the highway mattered to the local economy. For example, the *Rome Tribune-Herald* published a special fifty-four-page "Good Roads" edition on May 31, 1916. Boasting "that all roads lead to Rome," the editors of the newspaper claimed, "Georgia has outstripped any other state traversed by the Dixie Highway. From Macon to the Florida line it is now possible to travel over a good road. The Eastern Division via Savannah and Brunswick has only to build two bridges to solve the problem of [a] Florida crossing and provide tourists with a wonderfully fine drive to Jacksonville."[42]

A substantial amount of support for the Dixie Highway also came from civic groups in Georgia. For example the Rome Rotary Club passed a resolution in 1924 to "enlist the support of other Rotary Clubs and civic organizations, especially along the route, to make the Dixie Highway a paved way from the Great Lakes to the Gulf of Mexico."[43] Rome boosters also raised enough money to erect a pair of arches over the route at the boundary lines of Floyd County to greet automobile tourists.[44]

Another way that Georgians would promote the Dixie Highway was through the various businesses they built that catered to the tourists driving along the road. Even as early as 1916 the *Rome Tribune-Herald* had noted that "a passable highway between Rome and Atlanta . . . would place Rome in an ideal position to profit from the growing tide of Northern tourist traffic flowing to and from Florida."[45] The city soon had several filling stations, and by 1925 Rome had established a municipal camp for overnight automobile tourists at the American Legion Memorial Park on Shorter Avenue. In August of that year, cars with license plates from sixteen states were found in downtown Rome. The following month 913 automobiles carrying over twenty-one hundred tourists had stopped to stay at the camp.[46]

That communities along the Eastern and Western divisions of the Dixie in North Georgia competed for a share of the tourist trade on the highway is certain.[47] One indicator of the competition for traffic was the staging of motorcades that brought the visitors to towns along the Dixie Highway. In the fall of 1915—just six months after the formation of the Dixie Highway Association—the directors staged a motorcade between Chicago and Miami. According to the headline of a front-page article in the October 15, 1915, edition of the *Atlanta Constitution*, "Dixie Highway Tourists Will Arrive in Atlanta Late This Afternoon; Old Fashioned Southern Welcome Will Be Given Visitors to Georgia."[48]

Other motorcades were held on both divisions of the highway during the 1920s; these included trips on the Eastern Division between Chattanooga and Atlanta, which passed through Dalton and Calhoun.[49] Apparently the motorcades were as much to advance road improvement efforts as they were to promote the commerce that would occur in the towns along the Dixie's routes (fig. 3.5).

FIG. 3.5. Dixie Highway Motorcade in Calhoun, Georgia. Courtesy of Georgia Archives.

In terms of the commercial contest between the north Georgia towns along the two divisions of the Dixie Highway, the winner emerged prior to World War II. A review of a series of 1940 highway maps for Catoosa, Walker, Floyd, and Whitfield counties clearly indicates that the Eastern Division had many more of roadside businesses. In Catoosa and Whitfield counties, thirty-nine gas stations and eleven tourist camps stood along the rural stretches of U.S. 41. On the portion of U.S. 27 running through Walker and Floyd counties, just six gas stations and four tourist camps lined the highway in rural areas, and no such businesses stood along U.S. 27 in Catoosa County.[50]

A comparison of two North Georgia counties serves as an example. Whitfield and Floyd counties are largely similar by virtue of each having a large industrial town within its borders—Dalton in Whitfield and Rome in Floyd. In the case of the first county, twenty-seven gas stations and six tourist camps along U.S. 41 appear on the 1940 map; conversely the map produced the same year for Floyd County shows only three gas stations and just one tourist camp near the corridor of U.S. 27. Because the number of businesses was tied to the number of tourists who used the routes, it seems likely that more travelers preferred U.S. 41 to U.S. 27. Hence the large concentration of roadside buildings standing next to U.S. 41 and the sparse collection of similar resources on U.S. 27 as shown by these maps suggest that the Eastern Division had by 1940 eclipsed the Western Division in terms of the number of travelers who drove the highway in North Georgia.[51]

The preference that the motoring public had for the Eastern Division may also have something to do with the mileage of the two branches. Compared to the Western Division, the Eastern was a shorter and more direct route between Chattanooga and Atlanta: 126 miles versus 139 miles.[52] Another reason may have been that following the infusion of federal dollars into highway projects after 1921, the state directed its efforts at improving the Eastern Division sooner than it did the Western Division. Of the two explanations—the shorter distance—seems to be the more likely reason. A glance at a current Georgia highway map makes this factor apparent in that Interstate 75, the modern access-controlled highway between Chattanooga and Atlanta, closely parallels U.S. 41 for much of the nearly 115 miles between the two cities. At the same time, U.S. 27 is largely distant from Interstate 75.[53]

As for the other counties that the Dixie Highway's two routes passed through, it is difficult to gauge which branch had greater importance. Aside from the North Georgia county maps, none has been found showing the location of roadside businesses in the remainder of the state. It is likely, however, that residents all over Georgia were putting up gas stations, tourist camps, restaurants, and roadside businesses near the Dixie's two divisions soon after their routes were announced (fig. 3.6). One well-known example, which began in Eastman on U.S. 341, is Stuckey's Pecan Shoppes.[54] The Dixie Highway was also a conduit for outside influences that would do much to alter the region's uniqueness.[55] Soon the latest examples of modernistic architecture were found in Georgia towns along the Dixie's routes: Roy's Grill in Rossville; Dalton's Wink Theater; Milledgeville's Southern Stages Bus Station; the White House Restaurant in Savannah; Albany's Lee Drug Store; and the Waycross Restaurant in Waycross.

Savvy Georgia entrepreneurs made the connection between the Dixie Highway and their businesses even more apparent by adopting the name "Dixie" in the names of their enterprises. For instance C. S. Little of Waycross called his business along

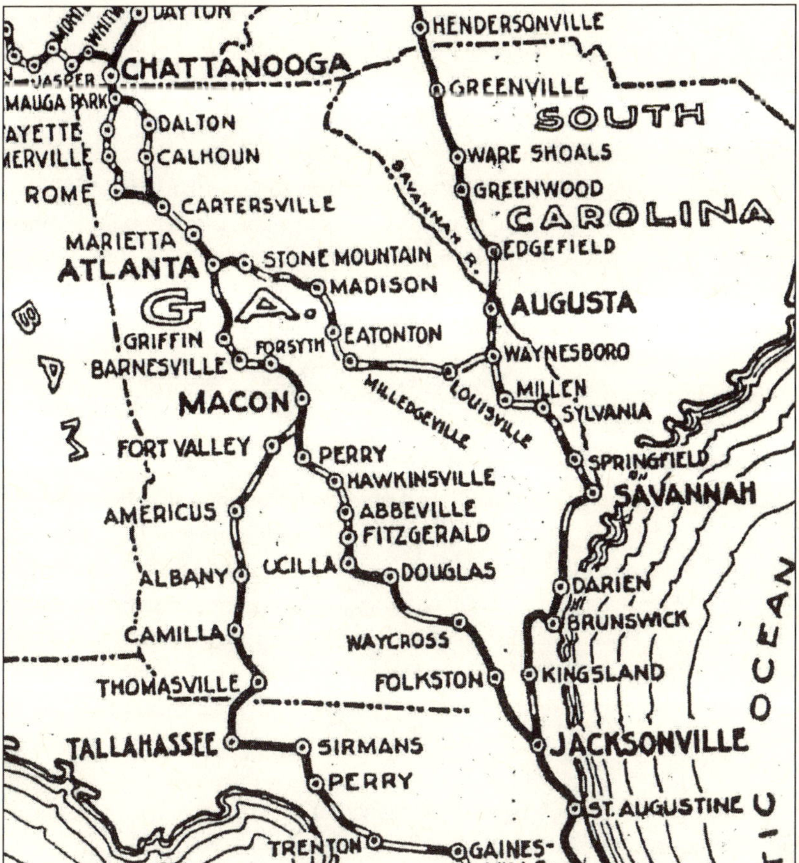

FIG. 3.6. Map of the Dixie Highway through Georgia. From the collection of Martha Carver.

the highway south of town "Dixie Camp." While their numbers are diminishing, other roadside establishments serve as reminders of the Dixie Highway's significance to Georgia: the Dixie Motel north of Ringgold; Camp Dixie near Tunnel Hill; Dixie Car Wash and Dixie Auto Sales in Dalton; Dixie Cab in Forest Park; Augusta's Dixie Court; Dixie Motel in Adel; Valdosta's Dixie Court; the Dixie Pine Motor Court in Bainbridge; and the chain of Dixie gas stations still found in some Georgia towns.[56]

Apart from instigating some of the crass boosterism and commercialism that came with advocating good roads, Georgians would play other significant roles in the Dixie Highway Association. For example, the Peach State hosted the organization during the meetings of its board of directors on four separate occasions: at the Hotel Dempsey in Macon on March 25, 1916; at the Hotel De Soto in Savannah on May 25 and 26, 1924; and at the Hotel Forrest in Rome on November 14 and 15, 1924, and again on November 15, 1925.[57]

By November 1924 the Dixie Highway Association had nineteen chapters representing 803 members. Women could not belong to the larger Dixie Highway organization, but they could form auxiliary groups.[58] The auxiliaries helped beautify the Dixie Highway, and by 1922 efforts were underway to make the highway a "Road of Remembrance" to World War I veterans.[59] A result of this latter effort is the American Legion Gate on Rossville Boulevard/U.S. 27 south of Chattanooga, Tennessee.[60] The earliest record of a local auxiliary of the highway association is the March 1916 formation of the Bibb County Women's Dixie Highway Auxiliary, with several Macon women serving as officers.[61] Perhaps the influence these women wielded in the organization led to the first meeting of the National Women's Auxiliary, held May 26 and 27, 1924, in Savannah.[62]

Despite the disbandment of the highway association in 1927 and the calamitous economic conditions of the Great Depression, work on the completion of a hard-surfaced roadway for the Dixie Highway continued during the late 1920s and the 1930s. A 1943 map compiled by the Public Roads Administration of the Federal Works Agency indicates that much progress had occurred in paving the route of the Dixie in North Georgia. This map suggests that the Eastern Division had achieved supremacy over the Western Division, because it shows that the Eastern Division had a continuous stretch of paved road between Atlanta and Chattanooga, but the entire segment of the Western Division running through Chattooga County was shown as having no hard surfacing.[63]

That the Dixie Highway represents a significant development in early-twentieth-century history suggests that it is also important for preservation. Unfortunately interest in the preservation of historic automobile highway corridors is still a recent phenomenon and has been largely confined to places outside the South. A trip on many roads that once carried the Dixie's traffic shows that much of the route has been destroyed or heavily altered. Yet portions of the highway and roadside buildings remain even now.[64] These intact stretches of roadway provide a fleeting glimpse of the way—literally as well as figuratively—earlier Americans traveled during the early twentieth century. How much the view of that time is unimpaired may determine to what degree these segments are actively protected. How many of the physical reminders are preserved will affect our ability to retrace the Dixie Highway without becoming sidetracked.

NOTES

1. Dixie Highway Association, "Minutes" (1915): 1. These documents are part of the collection found in the Chattanooga Automobile Club papers at the Bicentennial Public Library, Chattanooga, Tennessee.

2. Howard Lawrence Preston, *Dirt Roads to Dixie: Accessibility and Modernization in the South, 1885–1935* (Knoxville: U of Tennessee P, 1989), 52–55. Preston contends that the region lagged far behind the North in the ability to improve roads (12–13, 95–97). However, the quality of roads was a national problem. For example, see the account of Emily Post's trip west of Chicago on the Lincoln Highway in Drake Hokanson, *The Lincoln Highway: Main Street across America* (Iowa City: U of Iowa P, 1988), 29–30.
3. *Dixie Highway* 10 (May 1924): 10–11; *Georgia Highway Map* (Atlanta: Georgia Department of Transportaion, 1998–99). Geography had much to do with why the state would have such a large amount of the highway's mileage. With a total area of 58,060 square miles, Georgia is the largest state east of the Mississippi River and ranks twenty-first out of the fifty states in land area. Just four other states along the Dixie Highway (Michigan, Kentucky, Tennessee, and Florida) had both of the two major routes of the highway crossing them.
4. Preston, *Dirt Roads to Dixie,* 56–58; *Dixie Highway* 10 (May 1924): 10–11.
5. Dixie Highway Association, "Minutes" (1925): 10; *Dixie Highway* (May 1924): 10–11; Georgia Highway Map, 1998–99. The Dixie Highway "Minutes" usually refer to the Dixie Highway as a division, such as "Western Division," capitalized. The *Dixie Highway* magazine and minutes sometimes refer to it as a branch, such as the western branch. For consistency, the author has primarily used "Division," capitalized.
6. Chickamauga-Chattanooga National Military Park is one of the three units of the National Park Service associated with the Civil War that lie along the former routes of the Dixie Highway; the other two are Stones River National Battlefield in Murfreesboro, Tennessee, and Kennesaw Mountain Battlefield Park near Marietta, Georgia.
7. Dixie Highway Association, "Minutes" (1915): 10; *Dixie Highway* 10 (May 1924): 10–11; *Georgia Highway Map,* 1998–99; author interview with National Park Service historian Jim Ogden, Chickamauga-Chattanooga National Military Park, August 24, 1998.
8. Dixie Highway Association, "Minutes" (1915): 10; *Dixie Highway* 10 (May 1924): 10–11; *Georgia Highway Map,* 1998–99; Fort Oglethorpe U.S. Geological Survey Quadrangle Map, 1942; Estelle U.S. Geological Survey Quadrangle Map, 1947. The last two sources are located in the Map Collection, Science Library, U of Georgia, Athens, GA.
9. Dixie Highway Association, "Minutes" (1915): 10, 60.
10. Ibid; *Dixie Highway* 10 (May 1924): 10–11; *Georgia Highway Map,* 1998–99.
11. *Georgia Highway Map,* 1998–99; State Bridge Survey, Historic Preservation Division, Georgia Department of Natural Resources, Atlanta.
12. *Georgia Highway Map,* 1998–99; State Bridge Survey.
13. *Dixie Highway* 5 (March 1921): 7.
14. Dixie Highway Association, "Minutes" (1915): 10; *Dixie Highway* 10 (May 1924): 10–11; *Georgia Highway Map,* 1998–99.
15. Dixie Highway Association, "Minutes" (1915): 10; *Dixie Highway* 10 (May 1924): 10–11; *Georgia Highway Map,* 1998–99. The total mileage of the Dixie Highway's Western Division through Georgia was approximately 412 miles.
16. *Dixie Highway* 10 (May 1924): 10–11; *Rand McNally Auto Road Atlas of the United States,* facsimile of 1926 edition (Chicago: Rand McNally and Company, 1974), 26–27; *Georgia Highway Map,* 1998–99.

17. 1946 East Ridge U.S. Geological Survey Quadrangle Map, Map Collection, Science Library, U of Georgia, Athens, GA.

18. Apparently an earlier automobile route using the Johnston-Sherman name already existed prior to the creation of the Dixie Highway Association. According to an *Atlanta Constitution* article (March 23, 1915, 8), "It is the plan of the local association to have the old Johnston-Sherman highway made a part of the route and to this end pamphlets are being prepared giving the points of historical interest along the route."

19. *Georgia, a Guide to Its Towns and Countryside* (Athens: U of Georgia P, 1940), 300; Jim Miles, *Fields of Glory* (Nashville: Rutledge Hill P, 1989), 10; *Georgia Historic Bridge Survey* (Atlanta: Georgia Department of Transportation, 1998), 64; *Travaho Trails through Georgia* (Atlanta: Tourist Guides, 1946), 9; scrapbook of Atlanta newspaper articles (unpublished and no date), on file at Historic Preservation Division, Georgia Department of Natural Resources, Atlanta (hereafter cited as "Atlanta Scrapbook").

20. *Dixie Highway* 10 (May 1924): 10–11; *Georgia Highway Map,* 1998–99.

21. *Dixie Highway* 10 (May 1924): 10–11; *Georgia Highway Map,* 1998–99.

22. *Dixie Highway* 10 (May 1924): 10–11; *Georgia Highway Map,* 1998–99.

23. Dixie Highway Association, "Minutes" (1915): 10; *Georgia Highway Map,* 1998–99.

24. *Dixie Highway* 3 (April 1917): 1.

25. *Dixie Highway* 3 (April 1917): 13; Dixie Highway Association, "Minutes" (1916): 31–32; *Georgia Highway Map,* 1998–99.

26. *Dixie Highway* 3 (April 1917): 13; Dixie Highway Association, "Minutes" (1916): 31–32; *Georgia Highway Map,* 1998–99.

27. *Dixie Highway* 10 (May 1924): 10–11; *Georgia Highway Map,* 1998–99; John Gibson, "U.S. 17 Became Backbone of Camden County," *Southeast Georgian,* December 28, 1994, 5, 16; and Marsha A. Chance and Greg C. Smith, "Section 106 Consultation Case Report and Supporting Documentation for the St. Marys River Bridge Replacement Project Nassau County, Florida and Camden County, Georgia" (1993), 12–13 (copy of report on file at Historic Preservation Division, Georgia Department of Natural Resources, Atlanta).

28. Dixie Highway Association, "Minutes" (1916): 36–37.

29. Ibid.

30. *Dixie Highway* 10 (May 1924): 10–11; Georgia Highway Map, 1998–99. The total length of the Dixie Highway's Eastern Division through Georgia in 1924 was approximately 418 miles.

31. *Dixie Highway* 10 (May 1924): 10–11; *Georgia Highway Map,* 1998–99. The 1924 length of the Dixie Highway's Carolina Division through Georgia was approximately 268 miles.

32. Dixie Highway Association, "Minutes" (1916): 32; *Dixie Highway* 10 (May 1924): 10–11.

33. Dixie Highway Association, "Minutes" (1922): 106; *Georgia Highway Map,* 1998–99.

34. For example, a circa-1950 strip map of the Dixie Highway shows the highway largely following a single alignment on the route of the Eastern Division in Ohio, Kentucky, and Tennessee but entering Georgia north of Chatsworth on U.S. 411, which was never a town on the route adopted by the Dixie Highway Association during its twelve-year existence. Strip map in collection of Martha Carver.

35. Dixie Highway Association, "Minutes" (1916): 36. During the July 1, 1916, meeting of the Highway Association's Executive Committee, it resolved that the county had until January 1 to construct the road or the Dixie designation would be lost. Bartow County appears to have made good on fulfilling its obligation to maintain its segment of the Western Division route, and the alignment never was changed.

36. From 1920 to 1932, Georgia had 161 counties, but since 1932 it has had 159 counties, following the disbandment of Campbell and Milton counties.

37. Hokanson, *The Lincoln Highway,* 21.

38. According to the article "Cincinnati after Big Dixie Highway," *Atlanta Constitution* (March 31, 1915, 1), Howell was to give an address, "Newspapers Work in Building the Dixie Highway," at the Chattanooga governor's conference.

39. The eleven *Constitution* articles include "Great Good Roads Meet at Chattanooga in April to Be an Epochal Gathering," March 7, 1915, 3F; "Hoosierland-to-Dixie Highway Given Boost by Padington," March 18, 1915, 4; "Auto Men Enthuse over New Highway," March 23, 1915, 8; "Many to Attend Good Roads Meet," March 24, 1915, p. 7; "Atlantans Plan Chattanooga Trip," March 25, 1915, 7; "Many Georgians to Attend Convention in Chattanooga," March 27, 1915, 4; "Good Roads Clans and Developers from Illinois to Florida Preparing for Epochal Meeting in Chattanooga," March 28, 1915, 5; "Gordon County Will Boost Dixie Highway," March 30, 1915, 5; "Two Routes Are Proposed for Highway in Tennessee," March 30, 1915, 7; "Cincinnati after Big Dixie Highway," March 31, 1915, 1; and "Three Counties Pledge Roads for Highway," March 31, 1915, 7. The *Constitution* editorial "Work of a Master Builder" appeared in the March 26, 1915, edition, 8. In addition to the articles and the editorial, an advertisement, "Dixie Highway Convention Chattanooga," appeared in the March 31, 1915, issue, 7, where two railroads—the Western and Atlantic Railroad and the Nashville, Chattanooga and St. Louis Railway—offered a special round-trip rate of $4.35 on a charter train for those interested in attending the April 3 meeting.

40. "Hoosierland–to–Dixie Highway Given Boost by Padington," *Atlanta Constitution,* March 18, 1915, 4.

41. *Atlanta Constitution*, October 15, 1915, 1.

42. Roger Aycock, *All Roads to Rome* (Roswell, GA: W. H. Wolfe Associates, 1981), 300–301.

43. *Dixie Highway* 10 (May 1924): 19.

44. *Dixie Highway* 10 (February 1924): 15.

45. Aycock, *All Roads to Rome,* 301.

46. Ibid., 303, 305–6, 384.

47. Articles appearing in the *Atlanta Constitution* indicate that this competition appears to have begun even before the April 3, 1915, governors conference convened and before the Dixie Highway Association's selection of a route. See articles "Good Roads Clans and Developers from Illinois to Florida Preparing for Epochal Meeting in Chattanooga," *Atlanta Constitution,* March 28, 1915, 5; "Gordon County Will Boost Dixie Highway," *Atlanta Constitution,* March 30, 1915, 5; and "3 Counties Pledge Roads for Highway," *Atlanta Constitution,* March 31, 1915, 7.

48. *Atlanta Constitution,* October 15, 1915, 1.
49. Sherry Konter, *Vanishing Georgia* (Athens: U of Georgia P, 1982), 170, 195–196; Atlanta Scrapbook, "Jaw Cracker," 3.
50. Georgia State Highway Board Maps for Catoosa, Floyd, Walker, and Whitfield counties, 1940.
51. Georgia State Highway Board Maps for Floyd and Whitfield counties, 1940.
52. *Dixie Highway* 10 (May 1924): 10–11; *Georgia Highway Map,* 1998–99.
53. *Georgia Highway Map,* 1998–99.
54. For a discussion of Stuckey's, see Lisa Raflo and Jeffrey Durbin, "Teal Roofs and Pecan Logs: A History of Stuckey's Pecan Shoppes," *SCA Journal* 13 (Fall 1995): 2–8; Elizabeth McCants Drinnon, *Stuckey: The Biography of Williamson Sylvester Stuckey* (Macon, GA: Mercer UP, 1997); and Tim Hollis, *Dixie before Disney: 100 Years of Roadside Fun* (Jackson: UP of Mississippi, 1999), 21–41.
55. For a discussion of the cultural and social changes that resulted from the Good Roads movement in the South, see chapters 6 and 7 of Preston, *Dirt Roads to Dixie.*
56. Postcards from author's collection; author interview with Gretchen Kinnard, September 18, 1998, Atlanta, Georgia; United Motor Courts 1942 Directory, 28 (author's collection).
57. Dixie Highway Association, "Minutes" (1916): 30; (1924): 121, 149, 175.
58. Ibid. (1925): 175.
59. For more information about the Road of Remembrance Program, see Brian Butko, *Pennsylvania Traveler's Guide: The Lincoln Highway* (Mechanicsburg, PA: Stackpole Books, 1996), 111–13. In addition, see Thomas Shlereth, *U.S. 40: A Roadscape of the American Experience* (Indianapolis: Indiana Historical Society, 1985), 72–74, for a discussion of the importance of monuments along highway routes.
60. Georgia Bridge Survey, 63; Dixie Highway Association, "Minutes" (1923): 118.
61. Dixie Highway Association, "Minutes" (1916): 30–33; *Dixie Highway* 10 (May 1924): 16–17.
62. Dixie Highway Association, "Minutes" (1925): 176.
63. Federal Works Agency, *Highway Map of North Georgia,* 1943. U.S. Geological Survey Quadrangle Maps of North Georgia from the 1940s also seem to suggest that the Eastern Division had become the more important of the two divisions because U.S. 41 is labeled as "Dixie Highway," but the route of the Western Division, U.S. 27, shows no such name (see USGS Quadrangle Maps for East Ridge, Ringgold, Tunnel Hill, Fort Oglethorpe, and Estelle). Map Collection, Science Library, U of Georgia, Athens, GA.
64. Peirce Lewis refers to such collections of objects as some of the "innumerable displays in America's ever-expanding outdoor museum of discarded geographic artifacts." See Lewis, "The Landscapes of Mobility," in *The National Road*, ed. Karl Raitz (Baltimore: Johns Hopkins UP, 1996), 44.

CHAPTER 4

PLEASING THE EYE: BRICK PAVING AND THE DIXIE HIGHWAY IN THE SUNSHINE STATE

WALTER S. MARDER

The history of the Dixie Highway and how it helped open up the South for development cannot be written without mentioning Carl Fisher. Conversely, few would think of Fisher when writing the history of construction materials used for the Dixie Highway. Yet Fisher's use of brick at his Indianapolis Speedway showed how successful a paving material vitreous brick was for roadways. The state of Florida seemed to agree and built much of its section of the Dixie with brick. In the early part of the twentieth century brick was the "best road south," but the popularity of brick was short-lived. Less expensive and more readily available materials soon super-ceded it for paving roads. Walter S. Marder discusses the rise and fall of brick as a paving material in Florida.

On a hot, sultry Florida day, as only a day in Florida can be, men went to work laying bricks on a graded road to fill in another blank in what would become the famed Dixie Highway. Rain showers came and went, and still they worked on, until, at the end of the day, piercing tiredness left them with little energy except for sleep. The next day it began again. And, yet, what they created is one of the most lasting legacies of the great Florida boom: a road for tin can tourists leading literally to the Sunshine State and figuratively to a tourism industry that is still Florida's mainstay.

The Dixie Highway played a major role in the growth of Florida. There were several sections of the highway that were constructed of brick, a few of which remain today. They represent the only portions left of the Dixie Highway in Florida that are in their original state and which can be driven on today as they were so many years ago. The charm of what is left of that brickwork remains a legacy of those early laborers of Florida road building. Those brick roads are actually pleasant to drive on; because one must drive a little more slowly or destroy shock absorbers, they remind us too of a time when life was not quite so hectic as it is now.

Driving into Florida from Georgia, one suddenly senses a change. It is subtle but definite, as, along the area's east coast, the saw grass and broad rivers are left behind, and the sandy scrub, palm trees, and flocks of snowy egrets pick up. The traveler notices a tropical ambience deep in the kind of Dixie in which Miami Beach developer Carl Fisher wanted Americans to believe. The idea was not new with Fisher; Henry Flagler had done the same thing with railroads a generation earlier. Arriving in Jacksonville on the Eastern Division of the Dixie Highway, one spies the sparkling St. Johns River and then heads south toward the coast, eventually reaching Miami Beach or Tampa. As far south as New Smyrna, this has always been the case, ever since the English built their road south from Savannah to the colony of east Florida. Now on Interstates 295 and 95, and Interstate 75 on the west, the traffic is fast, and the scenery goes by too quickly to really appreciate or enjoy. The Western Division of the Dixie Highway took travelers through the state's high central ridge before veering off to the coast at Tampa. It is a different sort of scenery, more like the rolling hills of Virginia and the piedmont region of the Carolinas.

It was not that long ago that the Dixie Highway wended its way along or close to these same routes. While a somewhat perilous drive, as it is today, the dangers came not from heavy traffic or drivers going ninety miles per hour, but from ruts and mud and stumps in the road. But overall it likely was more pleasurable. There are still old Dixie Highway exits all along the course of I-95, well worth pursuing—a little slower perhaps, but infinitely nicer than the interstate. The traveler who takes this older route is in awe of the beauty of the pinelands and rivers lined with sabal palms. A lasting part too of the charm of that old Dixie Highway are the few remaining brick sections, a relatively novel building material at the time, but one that has far outlasted its competitors.

Brick as a paving began being used in 1893 in Ohio.[1] Florida's first brick streets were constructed in 1903.[2] Its first brick highway, constructed in 1912, was a part of the then-unnamed route between Jacksonville and Tampa. By 1915, though, the state had a road department, which would dramatically increase the Florida highway system.[3] It soon became apparent that the many reasons given for building good roads, the post, commerce, and agriculture, all paled in view of the benefit to tourism that these roads would provide. Florida's concerns were the road between Jacksonville and Pensacola, eventually called the "Old Spanish Trail," and routes from north to south, one from Jacksonville to Tampa and the southwest coast and one down the east coast of the state to Miami. The system eventually also added a route through central Georgia down Florida's west coast. It was at this time that the Dixie Highway movement began, which became a prime focus for Florida's north to south efforts. Its completion created a major network of roadways.

Road building in Florida did not start with the Dixie Highway, though. The first major road in Florida was constructed not by the Spanish, but by the British, between 1763 and 1775. It consisted of a graded roadway from the Georgia border to Jacksonville to St. Augustine and thence to Pellicer Creek, Bulow Plantation, and the Tomoka River, ending in New Smyrna, effectively linking all of the established British communities. It was sixteen feet wide.[4]

The making of highways for commerce and military purposes is ancient. The Roman Empire developed a sophisticated system of primary and secondary roadways: the primary for strategic purposes, the secondary for commercial purposes. By medieval times, road building had fallen off dramatically. The old Roman roads continued to be used, but few new roads were built, and maintenance was sporadic. The repairs and the construction of what few new roads were being built were financed in part by a road tax. It required able-bodied members of the community to do actual labor on the roads. This practice lasted well into the nineteenth century, both in Europe and in the United States.

The eighteenth and nineteenth centuries brought new improvements to road construction techniques. The *Builder's Dictionary,* published in London in 1734, defined "pavement," in part, by stating that "In *Venice,* the Streets, &c. are pav'd with Brick"; and "In *Amsterdam* and the chief Cities of *Holland,* they call their Brick Pavement, the *Bourgher Masters Pavement,* to distinguish it from the Stone or Flint Pavement, which is usually in the middle of the Street, fearving for the Paffages of their Horfes, Carts, Coaches and other Carriages; the Brick Borders being defign'd for the Paffage of People on Foot."[5]

The nineteenth century saw what some thought was going to be a substitute for roads altogether: the development of the railroad. It seemed unnecessary to have major trunk roads when railroads could be built that were all-weather, reliable, and fast. For many years thereafter, roads were again neglected, as they had been a thousand years before. Slowly, however, it was seen that roads were still needed to get goods and produce to and from railheads and depots. Thus road building was taken up anew to provide a means for U.S. farmers to ship their products. For most of these farmers, this was an everyday concern.

By 1892 the League of American Wheelman had formed the National League for Good Roads and in the following year, 1893, the first Good Roads convention was held in Washington, D.C. At the same time, the U.S. government set up the Office of Road Inquiry in the Department of Agriculture, to study the need for better roads. It was at about this same time that brick was being experimented with for the paving of rural highways. Brick had been used on a limited basis for paving city streets beginning in 1870, when the first such pavement was laid in Charleston, West Virginia.[6] In the fall of 1893 a brick roadway, considered the first such highway in the United States, was constructed in Ohio. While a similar project had been funded earlier in Illinois in 1892, that road was not opened until 1896.

The Ohio road was the Wooster Pike, running between Cleveland and Wooster. Four miles were paved with vitrified brick in a stretch of road between the towns of Parma and Albion. The brick part of the road was eight feet wide at one side of a total of twenty-four feet of roadway; the rest was graded and compacted. The brick was to be used in wet or snowy weather while the graded part was for summer use. The right-of-way was sixty feet wide, and the road cost sixteen thousand dollars per mile to construct.[7]

The Illinois brick road was opened in 1896, the first in that state, in the small town of Monmouth. The road was three thousand feet long and cost about three thousand dollars. While the money for the paving had been voted for the brick paving of city streets and rock roads, "These [rock roads] have proven very satisfactory, but the commissioners thought the brick road would be better and just as durable, so they tried the strip west of town as an experiment."[8] Unfortunately neither of these roads remains today.

In Florida, roads were regularly constructed of sand-clay; oyster shells, usually taken from ancient Indian middens; gravel; marl; limerock; or pine straw. Interestingly the pine-straw road was simply that, pine straw laid about a foot thick on the earth bed and allowed to be worn down and into the road surface. The straw roads cost about forty dollars per mile as compared to sand-clay's one thousand dollars per mile.[9]

In 1912 the federal government entered the road-building picture again under the aegis of the post office and enacted the Post Office Appropriations Act, enabling, for the first time, the use of federal funds for state road construction. Shortly thereafter, in 1916, the Federal Aid Road Act followed. That act, through the Office of Public Roads and Rural Engineering, Department of Agriculture, mandated that any state wishing to use federal funds, must have a state road department in place. It was this requirement that most effectively started Florida's road-building program.

While brick was being used increasingly for the surface of inner city and town streets, as a highway pavement, it continued to languish behind macadam. In 1909 another competitor to brick, concrete for roads, had been developed in rural Michigan. In use since 1865 in Scotland, concrete was still a novelty in the United States at the turn of the century. Occasionally used in the construction of buildings, it was most often employed for building foundations, for bridges, and as cast stone. Its ease of application, however, destined it for use on roads.

With money now available for the construction of roads, why did Florida choose brick as the material to use? What could possibly be its advantage over other materials? Brick gained use during the late nineteenth century for many reasons, one of which was cleanliness. Accustomed to dirt or sand streets that became quagmires in wet weather and were constantly fouled by the horses and mules using them, brick was seen by the public as an enormous improvement for paving. A 1908 publication extolled the virtues of brick roads, saying,

> The advantages of brick pavements may be stated as follows:
>
> (1) Ease of traction.
> (2) Good foothold for horses.
> (3) Not disagreeably noisy.
> (4) Yields but little dust and mud.
> (5) Adapted to all grades.
> (6) Easily repaired.
> (7) Easily cleaned.
> (8) But slightly absorbent.
> (9) Pleasing to the eye.
> (10) Expeditiously laid.
> (11) Durable under moderate traffic.[10]

The only disadvantage seems to have been the occasional defective lot of bricks that had been made too porous or too soft.

The brick used for paving was vitrified, that is fired in a kiln to a point where the clay and sand actually melted and fused together throughout the brick itself. The brick was laid on a subsurface or base. The base was sometimes merely graded and compacted earth, but more often a concrete or brick base was used. After the base was laid and properly sloped, the finish bricks were laid on top, usually in a bed of mastic such as asphalt. On completing the wearing surface, the interstices between bricks were filled with cement grout. Often the bricks were manufactured with small raised dots or lugs to keep them spaced properly from one another when being put down. The brick itself was generally about 3" x 4" x 9" and weighed about 9½ pounds, requiring forty-five bricks to the square yard of pavement.

FIG. 4.1. Western Division of the Dixie Highway (old Florida State Road 3) in Longwood, showing Robbins Brick paving. Photo by the author.

The colors were brilliant, ranging from deep purples through pinks, reds, and pale yellows, often with pronounced flashing. Examples include companies like Copeland and Inglis, Graves, and Southern Clay from Alabama and Augusta Block and Rockmart from Georgia. Some bricks came from Tennessee, such as those from Robbins Brick (fig. 4.1).[11]

In 1907 the cost of a yard of brick paving ranged from $1.30 to $2.60.[12] In 1903 Jacksonville had 11.2 miles of brick streets, or 233,912 square yards, a respectable amount but nowhere near Cleveland's 117.6 miles, or 2,016,349 square yards. Atlanta had only two miles and Augusta only one-tenth of a mile of brick paving. Brick was already becoming a Florida favorite.[13]

Brick paving achieved one of its earliest coups with the construction of the Indianapolis Speedway, in which Carl Fisher was involved. Started in 1909, the Speedway was paved with a macadam-like surface of tar and gravel that was then oiled. With great fanfare the track opened on August 9, with a 300-mile race. The paving began to break up badly as the day wore on, and vehicle after vehicle lost control on the rough surface. The resulting accidents caused the death of one driver, two riding mechanics, and two spectators and stopped the race before the 300 miles were finished. Chagrined Speedway officials quickly set to work to repave the entire track saying, "$100,000 will go to making the track a perfect one, as a new surface will have to be laid and this will be done in a thorough manner before cars will again be allowed to circle the track."[14]

The Speedway organizers were then approached by the National Paving Brick Manufacturers Association,[15] who proposed the use of a patented interlocking brick, Culver Block, for the new surface.[16] The decision was made to go ahead and pave with brick, "the finest material available at the time,"[17] but the bricks were grouted in place instead of using the interlocking system.[18] The net result of this for brick was a huge gain in reputation for its dependability as a paving material for high-speed roads.

The years from 1912 to 1916 were busy ones for road interests and the motoring public. The federal government's most intensive intervention into road building during that time occurred in 1913, in Florence, Alabama, when the first federally funded post road was constructed. The recently enacted Post Office Appropriations Bill promised each state ten thousand dollars, which was to be one-third of the cost of any new roads over which the post was carried; the state would fund the remaining two-thirds.

Brick highway paving finally came to Florida in 1911–12. The new road ran south from Jacksonville to present-day State Route 211 in Clay County. The roadway was fifteen feet wide and, from evidence remaining, constructed of Graves brick. It began in Jacksonville as New York Avenue, which eventually became the beginning of State Road 3. This was the start of the celebrated Dixie Highway, which would funnel millions of travelers into Florida over the years.[19]

Gov. Park Trammell authorized the first Florida State Road Department through Chapter 6883 of the Acts of 1915. At 3:30 P.M. on Friday, October 8, 1915, four men met in Tallahassee at the State Capitol for the purpose of organizing the department. The state would still not fund the construction of new roads. This cost was borne by each county through the issuance of bonds.

While the new department was getting started, the federal government was also busy at work. The momentum of the Good Roads Movement and the success of the post office roads project led to the decision in 1916 to enact the first genuine federal highways appropriations act. The Sixty-fourth Congress passed Chapter 241 July 11, 1916, noted as "An Act to provided that the United States shall aid the States in the construction of rural post roads, and for other purposes."[20] The language of the act had two requirements, that each state to be aided have a highway department in place and that none of the new roads was to be a toll road. Money was appropriated by the Treasury Department for this construction, starting with five million dollars for the fiscal year ending June 30, 1917. The next five years were also funded, increasing at a rate of an additional five million dollars per year, ending with an appropriation of twenty-five million dollars for the fiscal year ending June 30, 1921. The secretary of agriculture administered the act, and the money was to be spent on a fifty-fifty matching basis with the states. A true national Good Roads Movement was finally underway.

Meanwhile, in Tallahassee, William F. Cocke of Richmond, Virginia, had been hired as the first state road commissioner, and meetings were being held to determine the future of Florida's highways. On August 11, 1916, $112,000 was appropriated for 1917 and $116,000 for 1918 to meet Florida's federal share for those years. Because the department was not state funded until 1918, the appropriations were through the counties. Later the same month, the first state highway was designated: the road from Jacksonville to Pensacola.

In September 1916 Cocke made a trip to Hillsborough County on Florida's west coast for a "hasty inspection of the brick roads previously constructed."[21] One can assume that the reference was to brick streets in Tampa, as no recorded brick highways existed there at that time. He continued to Orange and Lake counties. In Orange County he made "an inspection . . . of the brick roads built under bind [*sic*] issue. These brick roads represent the highest type of highway work which had been done in the State up to the present time."[22] He was probably referring to those portions of the Dixie Highway that ran through Longwood and Maitland, both constructed of Robbins brick.

Orange County had begun its brick road movement as early as 1913. In that year a county commission resolution "propose[d] to construct these brick roads over the following routes: Oakland to Orlando via Winter Garden and Ocoee; Orlando to Zellwood via Apopka and Lockhart. From the Osceola County line to the Seminole County line via Taft, Pine Castle, Orlando, Winter Park and Maitland. Orlando east via Conway, at least five miles. Orlando east on Fort Christmas Roads, at least three miles."[23] Eventually, Orange County would authorize more than six hundred miles of brick roads.

Brick road news appeared regularly in Orange County during 1915 as the county prepared for its first brick highways. The section from Conway to Orlando was begun on March 31, and the road was finished by mid-May. It was then decided that the chairman of the county commission should become the county's delegate to the Dixie Highway convention to be held in Chattanooga on May 20, 1915. Orange County lobbied heavily to include the central part of the state in a western branch of the Dixie, and the *Orlando Morning Sentinel* of September 21, 1915, reported that such a route had been established.

The January 21, 1916, *Orlando Morning Sentinel* stated that work on the Dixie Highway between Orlando and the Osceola County line had been finished for nearly twenty miles. When that portion was completed, from Kissimmee to Apopka,

FIG. 4.2. Western Division of the Dixie Highway (old Florida State Road 2) in Kissimmee, Osceola County. Photo by the author.

it formed a continuous stretch of forty miles, the longest section of bricked roads on the western section of the Dixie highway[24] (fig. 4.2).

At the end of 1916, the chairman of the road commission called for a system of "Trunk Highways, which might be classed as Interstate Roads, with laterals across the State from the Gulf of Mexico to Georgia and from the Gulf to the Atlantic and connecting county seats and points of importance, forming a system of inter County roads" for "something must be done to relieve the aggravated condition which confronts travel into and out of the State."[25] The importance of tourism and convenience for the touring public had been recognized, and the chairman knew it had to be dealt with. Whatever other reasons there were for a good road system, tourists were now the most important.

Under the newly elected governor, Sidney Catts, a second road was proposed for north-south travel in 1917, to begin at the Georgia line and run through Jasper to Lake City. Additionally there was a call for an improved road from Jacksonville to Tampa. This part of the Dixie Highway eventually included a brick section in Putnam County, outside of Palatka and in Osceola County. In Osceola the recommendation was for a "Project between Kissimmee and Loughman, approximately 11 miles; brick surface 9 feet wide."[26]

The nine-foot road with shoulders, usually three or three and one-half feet wide and of clay or sometimes concrete, was widely accepted at this time (fig. 4.3). The concept of providing for two-way traffic would come later. In October 1917 the road department noted that

> with this object in view, the Department has conceived of two through lines of road as essential and that these two lines are recognized as trunk lines. One of these is the main highway across the State from East to West and the other through the length of the State North and South. . . . While it is the desire of the Department to secure the construction of broad surfaced highways for the main trunk lines, the Department will accept roads surfaced to a width of nine feet or more.[27]

The *Tampa Daily Times,* writing of the Dixie Highway from Jacksonville to Tampa and the need for major highways, quoted Good Highways Association president Bryan of Tampa: "If these highways had none but neighboring county connections, it would be difficult to estimate the development and business benefits to be derived from the county and city by their cooperation; but with the national connection coupled with interest of millions of tourists, investors and homeseekers in Florida, it is impossible to estimate what those benefits will amount to in a decade."[28] In the same article, the paper went on to write "the war will doubtless retard unimportant road work . . . but the building of connecting roads is necessary for prosecution of the war."

FIG. 4.3. A nine-foot-wide brick section of the Dixie Highway near Orlando in 1917. Florida Photo Archives.

In mid-1918 in Flagler County, the road south of Bunnell to the Volusia County line was ordered to be paved with brick. In Osceola County bricks were laid on the Kissimmee-Loughman Road, part of the Dixie Highway, up to Campbell Station. Within a few months, though, work would stall on this road as fuel to burn the bricks in Alabama was unavailable because of the war effort.

In September it was estimated that there were 473 miles of brick streets and highways in the state. The road department issued its *Second Biennial Report of the State Road Department for the Fiscal Year Beginning October 1, 1916 and Ending September 30, 1918* at the end of 1918. It reported on three brick roads being constructed. The Kissimmee to Loughman project, a nine-foot-wide road, had 8.9 miles constructed by the Alabama Paving Company; the Arcadia to Camp Dorr road, 10.2 miles of nine-foot-wide road, was completed by force account (convict) labor; and the Tarpon Springs to Clearwater Road had be constructed nine feet wide for twelve miles. The beginning of 1919 brought some specific data to light regarding the Kissimmee-Loughman portion of the Dixie Highway, as contained in table 4.1.

Completed brick	Sta 143 I 10 to Sta 218
Contract Price Estimate	$63,344.12
Amount estimated to Jan 1 '19	19,675.99
Balance to be completed	43,668.13
Brick laid to date	7450 sq. yds.
Brick on hand	26,000; will lay 580 sq. yds.
Curb on hand not laid	1000 lineal feet
Brick needed to complete work	
18,300 sq. yds. @ 42 brick per sq. yd.	768,600 bricks
Laying and grouting brick on hand	580 sq. yds.
@ 50c	$290.00
(Curb)	1000 lin. ft.
@ 10c	$100.00
(Grouting)	9000 cu.yds.
@ 35c	$3150.00
Additional Brick	18,300 Sq. yds. laid and grouted
@2.40	$43,920.00

TABLE 4.1. Roadbuilding data on Kissimmee–Loughman portion of Dixie Highway, 1919. From [Florida] State Road Department, "Minutes," February 7, 1919. Florida Department of Transportation, Tallahassee.

The road was not completed until 1920, after several starts and stops and rejected bids. In 1939 the WPA guide to Florida described this road in its Tour 2 (from Brunswick, Georgia to Tampa) with some flourish: "South of Kissimmee, the route follows a stretch of old brick highway, often inundated during heavy summer rains. During the construction of the road through the swamp in 1916–17, workmen supplemented their wages by catching baby alligators and selling them to curio shops."[29]

In May 1919, hotel owners from Daytona Beach, Ormond Beach, and other seaside resorts banded together and went to Tallahassee. There they petitioned the road department to shorten the coast route (by nineteen miles) from St. Augustine to Daytona. They wanted the section of the Dixie Highway designated as State Road 4 to be directed down the coast, instead of its route through Hastings and Bunnell. They eventually succeeded in this, and the still-existing brick road was cut out of that route.

May 1919 also saw a determination by the road department that would be the death knell of brick roads: a "committee to be vigilant in searching for new materials or, if necessary, new methods of road building."[30] Additionally the concept of the two-way road was broached for the first time. "Mr. S.N. Cornwall has been

FIG. 4.4. Orange crate label with traffic sign motif, from the Dixie Highway brand of the Leesburg Citrus Growers Association. Leesburg was on the Western Division of the Dixie Highway, also known as State Road 2. Photo by the author.

before the Department for the purpose of explaining his type of road, which is commonly known as the two-way road."[31] Cornwall's idea was at first accepted with the idea of construction of a test section, but that was soon abandoned. The two-way, sixteen-foot-wide road was, however, here to stay, thus further sealing the fate of the nine-foot brick road.

The year 1920 opened with a wider road: sixteen feet was now the official width, and the right-of-way went from fifty feet to sixty.[32] In 1920 there were eight million cars on the road in the nation, up from two million in 1915; this came to one car for every thirteen people. The press for more and better roads was increasing daily. The recently ended war had shown the value of truck traffic to move freight, and the optimism and economic upswing of the postwar years had everyone buying a car—and traveling.

FIG. 4.5. Orange crate label with road motif, from the Dixie Highway brand from Leesburg Citrus Growers Association. Photo by the author.

The postwar boom in Florida development even saw citrus growers getting into the highway act. There were at least three packing houses using labels that revolved around roads: Ward's Lake Charm Fruit Company in Oviedo had a "Cross Road" brand, and the Leesburg Citrus Growers Association had two "Dixie Highway" brands (figs. 4.4 and 4.5).

From 1922 on, however, little new brick roadway was constructed. Local materials, particularly limerock, had made their impact on road building and were dominating the field. In fact it was not long before sections of brick highways would be paved over with asphalt in order to make them wide enough for two-way traffic while other sections were abandoned entirely. The brick was not at fault; the roads simply had not been built for the new conditions.

Florida Highways, the departmental magazine, was started in 1923 as a way of demonstrating to the public the enormous strides that road building had made in Florida since the inception of the road department in 1915.[33] In February 1924 it reported that the state had constructed 24.896 miles of brick roads. This did not include that work funded by county bonding. The state had taken on the responsibility of funding some roads as a result of a 1921 federal law requiring that 7 percent of a state's roads be hard surfaced and meet federal standards. That 24.896 miles was the lowest figure of any type of work, with simple grading being the highest at 178.006 miles and bituminous macadam second at 138.77 miles.

The phenomenal Florida boom was in its heyday in 1925, and Charles Donald Fox, who had "just completed a tour of Florida" wrote a book called *The Truth about Florida*. Fox extolled the virtues of the Sunshine State, its agriculture and oranges, its beautiful beaches, its opportunities for "business and professional men," and its roads. He described the route between Daytona and Tampa as the "great million dollar Triangle Highway" and went on to say, "It passes through a beautifully wooded and lake region by way of DeLand, Sanford, Winter Park, Orlando, Haynes City, Lake Alfred and Lakeland. The first seven miles of this road is rock, then seventy-seven brick, asphalt block to Plant City, and the balance wide brick."[34] Seventy-seven miles of brick—an enormous accomplishment that today is represented by two stretches that together total less than one mile. In 1921, by consent of brick manufacturers around the country, the sixty-plus varieties of paving brick in use were reduced to a manageable six. The modern design of vitrified brick "provided a pavement not only of greater structural strength, more logical from an engineering standpoint and better suited to motorized traffic, but one that is also more economical in first cost."[35] The National Brick Paving Association went on in its praise: "Brick pavements are long-lived and, because of the hard, dense, tough units of which the pavement is composed, have shown their ability to resist the abrasive action of heavy concentrated fast moving traffic."[36] Brick pavements were said to have been in use for twenty years and more "with scarcely a brick having been disturbed."[37]

Ten years later, in 1931, in extolling vitrified brick as a paving material, a few years after brick paving had peaked,[38] the National Brick Paving Association declared that the only specification necessary for a vitrified paver was the "rattler" wear test, wherein iron shot was fired at the brick to break its surfaces and edges. "Only the strongest, most homogeneous, and toughest brick can withstand this test and those failing to meet it are rejected as paving material."[39] At brick's zenith, in 1926, Florida had 337 miles of rural brick highways, the third largest number in the United States, with Ohio first, at 1,412 miles and Pennsylvania second with 358.7 miles. Florida also had 389.5 miles of county and local brick roads in 1926, the second largest amount of any state.[40]

In 1926 the *Florida Engineer and Contractor* wrote in their March issue,

> What governed the selection of pavement for these roads (in Escambia County)? The fact that there was an abundance of good aggregates for concrete within the limits of the County and within a short rail haul of each project made concrete the

cheapest of the high type pavements. Portland cement was available from mills in Alabama. The specifications upon which bids were asked covered also brick and several types of asphalt paving, but long haul on these materials made the prices on them prohibitive in Escambia County. The only way by which they could have competed in price with the highest standard portland cement concrete pavement would have been to have lowered dangerously the standard specifications for them. The result was—portland cement paving throughout Escambia County.[41]

While both *Florida Highways* and the *Florida Engineer and Contractor* both carried many advertisements for concrete and asphalt products, neither had any ads for brick.

The *Fifth Biennial Report of the State Road Department,* put out on December 31, 1924, spoke of highway beautification programs, the need for better lighting, and the requirement for vehicles to be equipped with pneumatic tires. On the subject of road construction, it went on to state that for some years most materials used in the construction of roads had come from other states, but experiments were showing that Florida had an abundance of excellent road-building materials and local materials were being used throughout the state. Particularly limerock had been used in approximately 70 percent of the mileage of hard-surfaced roads constructed in 1923. By 1924 this figure was up to 90 percent. For many years limerock had been used as a base for brick, asphalt, and concrete highways. Now it had come into its own as a surface material. The day of the brick highway was over. In Osceola County the brick section of the Dixie Highway, which had been finished in 1920, was covered, except for one-half mile, in 1924.[42] The uncovered stretch still exists as Old Tampa Road in Intercession City.

The longest section of road still useable, though, is a stretch of eleven miles in Flagler County, abandoned long ago because a few hotel owners wanted the route changed (fig. 4.6). It is their lasting legacy. Those sections that are still there are almost in the original form, still there to be driven on with the sense of that time when they were the best to be had. Almost all of the old asphalt roads and concrete roads have been replaced and few traces remain. If careful, Florida has the chance to save the Dixie Highway, especially those original brick sections. While modern highways and freeways speed traffic through the state, the state has developed the Florida Scenic Highways program to focus on scenic and historic roadways that can enable this. Florida can save its brick roads so that they can become a permanent part of Florida's history and a reminder of the heady days when travel was an adventure shared on the Dixie Highway.

NOTES

1. Jean Labatat and Wheaton Lane, eds., *Highways in Our National Life, Symposium* (Princeton: Princeton UP, 1980), 85.

FIG. 4.6. Eastern Division of the Dixie Highway through Flagler County (old Florida State Route 104) between Bunnell and Espanola today and in a postcard view. Photo by the author; postcard from author's collection.

2. *St. Petersburg's Architectural and Historic Resources* (St. Petersburg: Community Development Department, Planning Division, 1981), 60
3. [Florida] State Road Department, "Minutes," October 8, 1915. The minutes of the departmental meetings cited here and below are archived at the Florida Department of Transportation, Tallahassee.

4. William R. Adams, Daniel L Schafter, Robert Steinbach, and Paul L. Weaver, *The King's Road: Florida's First Highway* (St. Augustine: Historic Property Associates, 1997), 11
5. A. Bettesworth and C. Hitch, *The Builder's Dictionary; or, Gentleman and Architect's Companion,* vol 2 (1734; reprint, Ottawa: Association for Preservation Technology, 1981), n.p.
6. Ira O. Baker, *A Treatise on Roads and Pavements* (New York: John Wiley and Sons, 1907), 462. Charleston experimented with less than one block in that year and went back in 1873 and paved a full block on a principal business street, using ordinary building brick; this street was still in service in 1907.
7. Albert C. Rose, *Historic American Highways* (Washington, DC: American Association of State Highway Officials, 1953), 98.
8. "The Brick Roads," *Monmouth* (Illinois) *Daily Record,* October 28, 1896, 1.
9. [Florida] State Geological Survey, *Bulletin* (1912). In 1912 Florida had 1,508.75 miles of sand-clay roads, 218.1 of shell, 26.5 miles of gravel, 846.8 miles of marl, 12 miles of limerock, and 273 miles of straw, with brick at a mere 8.5 miles, all in Duval County
10. Austin T. Byrne, *A Treatise on Highway Construction* (New York: John Wiley and Sons, 1908), 258.
11. Both Robbins and Coaldale were eventually bought out by Southern Clay.
12. Byrne, *A Treatise on Highway Construction,* 270.
13. Ibid.
14. "The Disasters at Indianapolis," *Literary Digest* 21 (September 11, 1901): 393.
15. The National Paving Brick Manufacturers Association was formed in 1905 to promote the use of brick as a paving material through technical information. The association changed its name to the National Paving Brick Association (NBPA), and then, in 1934, seven associations as well as the NPBA joined together to form the Structural Clay Products Institute, predecessor of the current Brick Institute of America. Conversation with Brian E. Trimble, staff engineer, Brick Institute of America, March 29, 1994. (Note: The Brick Institute of America has since changed its name to the Brick Industry Association.)
16. From a conversation with Donald Davidson, historian for the U.S. Auto Club in Indianapolis, March 29, 1994. The Culver Block had been patented in 1901, and more than 3 million bricks were used for the job, requiring the production facilities of seven different companies.
17. Jack C. Fox, *The Indianapolis 500* (Cleveland: World Publishing, 1966), 4–6.
18. Indianapolis was a brick course until 1938, when the first change was made, paving over the curves, which were considered too smooth. Finally in 1962 the home stretch was paved too, except for one yard left for sentimental reasons.
19. Lesta Mami, Florida Department of Transportation, informed this author that as with most of the state's other brick highways, New York Avenue was ultimately macadamized, but the brick had lasted until 1931. Personal conversation, March 17, 1994.
20. Sixty-fourth Congress, Session I, 1916: 355. Government Documents Collection, Florida State Library, Tallahassee.
21. [Florida] State Road Department, "Minutes," September 30, 1916.
22. Ibid.

23. Bradley Keeler, "Brick Highways in Orange County, Florida: A Report for the Orange County Historical Museum," unpublished manuscript, 1995, 7 (on file at Orange County Historical Museum, Orlando, FL).
24. Ibid., 12.
25. *Second Biennial Report of the State Road Department for the Fiscal Year Beginning October 1, 1916 and Ending September 30, 1918* (Tallahassee, 1918).
26. [Florida] State Road Department, "Minutes," July 11, 1917.
27. Ibid., October 10, 1917.
28. "One of the Great Cross State Highways," *Tampa Daily Times,* December 17, 1917, 1.
29. Federal Writers' Project for the Works Progress Administration for the State of Florida, *Florida: A Guide to the Southernmost State* (New York: Oxford UP, 1939), 6.
30. [Florida] State Road Department, "Minutes," May 8, 1919.
31. Ibid.
32. Ibid., January 29, 1920.
33. Curiously, under the circumstances, the August 1923 number started a format that would continue for some years: featuring a photograph of a brick highway at the beginning of the lead article.
34. This is also the Dixie Highway; Fox's use of the name "Triangle Highway" is the only such reference this author has seen. Charles Donald Fox, *The Truth about Florida* (New York: Charles Renard Corp., 1925), 143–44.
35. National Brick Paving Association, *Vitrified Brick as a Paving Material* (Washington, D.C., 1931), 4.
36. Ibid.
37. Ibid.
38. C. Harold McCollam, *The Brick and Tile Industry in Stark County 1809–1976* (Canton, Ohio: Stark County Historical Society, 1976). Metropolitan Paving Brick Company of Canton had its best year in 1923, shipping 92,000,000 pavers, or enough for 2,044,444 square yards—that's 681,481 miles of nine-foot-wide roadway. According to Brian Trimble, two years later, the U.S. brick industry produced the largest amount of brick pavers ever: 450,000,000
39. National Paving Brick Association, *Vitrified Brick,* 4.
40. U.S. Department of Agriculture, *Yearbook of Agriculture* (Washington, DC, 1926), 251.
41. Ibid.
42. Personal conversation with Lesta Mami of Florida Department of Transportation, March 23, 1994.

CHAPTER 5

CONNECTING THE EAST AND WEST COASTS: THE TAMIAMI TRAIL OF THE SUNSHINE STATE

CARRIE SCUPHOLM

For many automobile tourists the ultimate goal of traveling the Dixie Highway was to reach the new vacation paradise of Florida. For road-promoter Carl Fisher this north-south interstate highway would provide a route for northern vacationers to reach his recently completed resorts in Miami Beach at the southern terminus of the Dixie Highway. The Western Division of the Dixie Highway entered Florida near Tallahassee and followed a central inland route through the state, while the Eastern Division entered near Jacksonville and followed the Atlantic coastline, requiring several east-west connector routes (fig. 5.1). The Tamiami Trail, the road from Tampa to Miami, connected the east and west coasts of Florida. A portion of this connector route overlapped the Dixie Highway (from Punta Gorda to Miami), but the shared road was better known by its original name, the Tamiami Trail. After the mountains of Tennessee and the political infighting of Georgia, the Everglades of Florida presented a different challenge to completing the Dixie Highway. Once a road was constructed through the marshes and mangrove swamps, the Everglades developed into an attraction in and of itself. Completion of the Dixie Highway made "the land of sunshine" accessible to almost everyone. The Dixie Highway may no longer be the primary route to Florida, but tourists continue to travel south and constitute a sizable portion of Florida's economy.

> The Tamiami Trail is a unique highway . . . it took much more than money—more than average men, to pioneer, engineer, finance, gain public support and construct what was then considered by many impossible of success.
>
> **—Barron G. Collier Jr., "Markers on Trail Dedicated"**
>
> *Fort Myers News Press,* June 8, 1958

The Tamiami Trail links the two Florida cities for which it was named—Tampa and Miami. More commonly known as U.S. 41, the trail had a momentous impact on the state's growth. Francis W. Perry of Fort Myers and Capt. James F. Jaudon of Miami are credited with the initial idea for this road.[1] The 273-mile trail took some years to complete and cost more than $13 million; it opened to traffic April 25, 1928. An immediate success, this early highway united the lower east and west coasts of Florida for millions of motorists.

Although the Tamiami Trail is no longer the only route across the southernmost portion of Florida's peninsula, it continues to serve as a major business corridor along the southwest coast, linking towns and cities. The landscape it traverses ranges from commercial strips to planned developments to tourist attractions to Everglades swamp. For travelers who have the time, it still provides a scenic and leisurely route through the Everglades. Twenty-seven years after the road opened, the *Fort Myers News Press* reported the Tamiami Trail "has become so much a part of the everyday life of the communities along its route that the extraordinary features of its original construction are all but forgotten."[2]

EARLY HISTORY

The Tamiami Trail originated as part of the general road-building wave that swept over Florida in the mid-1910s. An automobile endurance race from Tampa to Jacksonville and back initiated the state's Good Roads Movement. A series of newspaper articles stressed the construction of good roads as essential to the material advancement of the state. According to the accounts, Florida could attract tourists for winter vacations if it could provide two main highways—one along the western Gulf of Mexico coast and one down the eastern Atlantic coast.[3]

In 1914 plans to extend the Dixie Highway southward along the coast from Jacksonville to Miami stirred interest in a "loop road" across the lower peninsula to connect with a west coast branch of the same highway. Newspapers on both sides of the state supported and widely publicized the idea. The determination of a location for the proposed highway produced considerable rivalry among the communities and real estate groups that stood to benefit by the surge in economic development and tourism sure to result from the new road.[4]

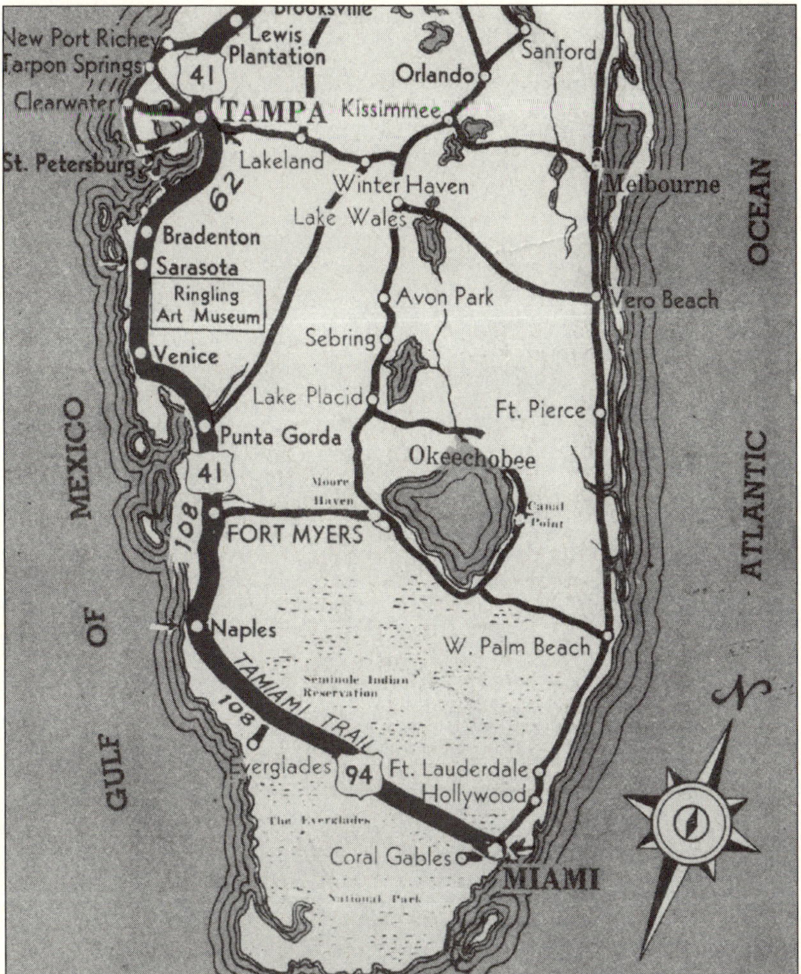

FIG. 5.1. The Tamiami Trail from Tampa to Miami connected the east and west coasts of Florida, as seen in this *detail from* a map showing the route of U.S. 41. From the author's collection.

Road boosters from central Florida attended the Dixie Highway Association convention held in Chattanooga, Tennessee, on May 20, 1915. They lobbied to have the new segment of the Dixie Highway (south of Jacksonville) constructed through central Florida. Although the effort to secure the route failed, this gathering led to the creation of the Central Florida Highway Association in June 1915. The first meeting in Orlando was probably the largest road convention ever held in Florida, attended by more than two thousand people. Here the association formally pledged to support several projects, including the Tamiami Trail. The Central Florida Highway Association was later described as "the greatest good roads booster association which has ever existed in Florida, responsible for the real beginning . . . of good roads in the central and southern portions of the state."[5]

FIG. 5.2. Sheet music for the Tamiami song. From the author's collection.

West	
Counties	***Their Towns***
Hillsborough	Tampa
Manatee	Bradenton
Sarasota	Sarasota
Charlotte	Port Charlotte
Lee	Fort Myers
Collier	Naples
Dade	Miami
East	

TABLE 5.1. Counties and towns along the Tamiami Trail.

In 1915 a group of businessmen from Miami, Naples, and Fort Myers met with state highway officials in Tallahassee to discuss the feasibility of a cross-state highway from Tampa to Miami via Bradenton, Sarasota, Venice, Fort Myers, and Naples. At first the proposed roadway was referred to as the "Miami to Marco Highway" and the "Atlantic to Gulf Boulevard." Conflicting accounts concerning the origin make it unclear as to when the name "Tamiami Trail" was officially adopted.[6] D. C. Gillett, president of the Tampa Board of Trade, and Francis W. Perry, president of the Fort Myers Chamber of Commerce, have each been credited with coining "Tamiami" from the combined names Tampa and Miami.[7] Local businessmen later formed the Tamiami Trail Association to promote and support construction of the highway.

The year 1915 also marked the creation of the Florida State Road Department by the state legislature. Following congressional passage of the Bankhead Act in 1916—which provided the first systematic federal aid for improving highways—the 1917 state legislature increased the road department's authority. The immediate goal was to supply hard-surfaced roads joining cities, towns, and villages (table 5.1).[8]

The segment of the route from Tampa to Fort Myers was constructed by the various counties prior to 1915, and only later was it designated part of the Tamiami Trail.[9] Before the Florida State Road Department was established, each county created a road and bridge district to finance road building within its boundaries. In 1911 Hillsborough County sponsored a bond issue to build a hard-surfaced road south to the Manatee County line. At the same time, voters in Manatee County approved a $250,000 bond issue to build a road from the Hillsborough County line to Sarasota. In 1914, through the efforts of the Sarasota Good Road Boosters, the Sarasota-Venice Road and Bridge District was created and financed $250,000 to continue the highway south to Venice. In Englewood citizens planned to build a road to Venice with the hope of becoming the southern link of the Tampa road system.[10]

Preliminary surveys for the cross-state segment of the Tamiami Trail through the Everglades were made in August 1915. The exact date when construction began on this segment is unknown, but the first payment for work performed was made in September 1916.[11] Yielding to the Tamiami Trail pressure groups, the Fort Myers District sold bonds worth $177,000, a fraction of the ultimate cost. Everglades City raised $125,000, intending to close the gap from Marco to the Dade County line. Dade County sponsored a bond issue for a road across the Everglades to the Lee

County line. Construction originated in 1917 but proceeded slowly because of the labor and material shortage created by World War I. Although the Dade County portion of the grading was complete in 1918, the Lee County portion of the trail was discontinued because of a lack of funds.[12]

In 1921 the fate of the cross-state segment of the Tamiami Trail appeared uncertain. Rumors circulated that this lower section would never be completed, especially not across the "impassable" Everglades. By 1923, with vast sums of money and materials expended, several workmen dead from drowning or dynamite explosions, and little progress made, south Florida residents doubted that the road could be completed. Then, in spring 1923, a group of public-spirited citizens calling themselves the Tamiami Trail Blazers set out to arouse public interest and demonstrate the feasibility of completing the highway.[13]

The Tamiami Trail Blazers attempted to navigate the thirty- or forty-mile stretch between the Dade County and Lee County ends of the trail, following the route previously forged by Capts. George Storer and J. F. Jaudon.[14] The trail-blazing expedition of ten cars, twenty-three men, and two Native American guides left Fort Myers on April 4, 1923. After a perilous three-week trip during which the group was reported lost several times, seven cars reached Miami, and the Tamiami Trail became the most discussed highway project in America.[15]

That same year, the state legislature created Collier County, largely a result of Barron G. Collier's promise to finance the final link of the trail if the state would turn his land holdings in Lee County into a separate county. Collier backed a $350,000 loan to relaunch construction and used his own construction company to do the work.[16] The Florida State Road Department officially recognized the project the following year. The state legislature incorporated the Tamiami Trail into the State Highway System and assumed responsibility for its completion.[17] Prior to this each county had been responsible for its own construction and maintenance. The absence of a coordinated system proved a major problem, hindering completion of the highway. To further complicate the situation, the road through the areas least developed and least able to finance themselves proved to be the most difficult and expensive to construct.[18]

Work on the road proceeded slowly. An average of 150 men working on the project were able to complete only 1.25 miles per month. The final 45.5 miles took nearly four years to build. Surveyors and men clearing the right-of-way worked breast deep in the swamps. Dynamite crumbled the hard rock under the swampy stretches east and south of Naples. Then a walking dredge moved in to rake up the boulders left by the dynamite, which were used as fill for the elevated road bed. A canal was formed along the north side of the alignment in the void left by the rocks.[19]

On April 24, 1928, the final link—a section of road from Naples to the Dade County line—was completed. The Tamiami Trail officially opened with ceremonies at Fort Myers on April 25, 1928, nearly thirteen years after the groundbreaking. To celebrate the completion of the highway, the Tamiami Trail Association arranged a county fair in Everglades City. Barron G. Collier led a huge motorcade that originated in Tampa and ended three days later in Miami. Captain Jaudon gave an address on the early history of the trail, and Gov. John W. Martin spoke on the value of the Tamiami Trail to the state.[20] A total of five thousand official Tamiami Trail flags were produced for the gala opening automobile caravan to be flown by participants in the motorcade; the day the trail opened, the *Fort Myers Tropical News* explained the flag's design:

> The flag is red, with a white band running diagonally across, and the Trail insignia is in the center. The red is symbolic of the uncharted wastes of the Everglades hitherto inhabited only by the redman. The wide band of white represents the Tamiami Trail penetrating this domain, the invasion of the white man. In the center of the band is the official monogram; being the initials of Tampa and Miami, with Tamiami Trail at top and bottom. The whole is enclosed by a double circle of black.[21]

If any of the flags survived, none are publicly known to exist.

Stone arches constructed by Collier County in 1928 once spanned the trail at two locations—the Lee-Collier County line and the Collier-Dade County line (fig. 5.3); they were demolished in 1956, when the trail was widened. Other road improvements included gasoline and rest stations built at ten-mile intervals in Collier County. In addition to providing gasoline and mechanical services, the stations also housed the Florida Mounted Police.

FIG. 5.3. Officials pose in front of the arch constructed on the trail between Collier and Dade counties. Courtesy of the Collier County Museum, Naples, Florida.

RECOGNIZING THE TRAIL'S SIGNIFICANCE

Recognition of the Tamiami Trail's historic significance occurred only thirty years after its completion. A June 1958 celebration of its thirtieth anniversary included the Historical Association of Southern Florida placing markers at three locations along the Tamiami Trail—Miami, Carnestown, and Fort Myers. As part of the celebration, association officials traveled by motorcade from marker to marker. The motorcade headed west from Miami, reversing the original path followed by the Tamiami Trail Blazers.[22] The *Collier County News* urged all county residents to "join in the commemorative events as a mark of respect for the courageous men who made possible this modern miracle of road building."[23]

In December 1998 the Florida Department of Transportation officially recognized a fifty-mile segment of the Tamiami Trail as a Florida Scenic Highway. The section runs from Collier-Seminole State Park to Dade County. The Tamiami Trail was the first highway in southwest Florida and only the second in the state to receive such a designation. The Tamiami Trail Corridor Advocacy Group, a grassroots organization composed of citizens, park representatives, and public planning officials, spent two years attaining this designation. This same segment of the Tamiami Trail was designated a National Scenic Byway by the Federal Highway Administration on June 15, 2000.

THE TRAIL TODAY

Florida roadside architecture has fared better than roadside architecture in many other states. Because so many tourists still drive to Florida, a significant number of establishments from the days of the two-lane road survive. As such, road-related architecture from the Tamiami Trail's heyday can be seen and visited today; the segment of the trail from Tampa to Fort Myers is older and more commercial than the stretch through the Everglades to Miami.

The Tamiami Trail (U.S. 41) commences at a nondescript intersection in Tampa just a few miles from Ybor City, the historic Cuban section of the city. Heading south, the highway winds away from Tampa Bay, past the Florida Tropical Fish Farms Association, and into Gibsonton, long a haven for retired carnival people. After Gibsonton the trail crosses farmland and proceeds to Ruskin. The landscape here is characterized by tomato-packing houses, empty billboards, and the occasional phosphate-mine hill. After Ruskin the trail continues into Bradenton's downtown historic district; south of here one encounters the first generic commercial strip—car dealerships, convenience stores, banks, liquor stores, video stores, motels, and a barrage of fast-food restaurants.

Proceeding on, it is difficult to determine where Bradenton ends and Sarasota begins, or where Sarasota ends and Venice begins. One of Sarasota's greatest cultural

FIG. 5.4. These 1930s postcard views highlight lush scenery and beautiful sunsets. From the author's collection.

attractions is located on the trail—the estate of John Ringling of the Ringling Bros. and Barnum and Bailey Circus. The John and Mable Ringling Museum of Art, opened in 1930, boasts one of the largest and finest collections of European paintings in the United States. Ringling's winter residence, circus museum, and a transplanted eighteenth-century Italian theater are also part of this sixty-eight-acre complex.

On the other side of Venice, one crosses the Charlotte County line, and the road opens to a panorama of pine trees and palmettos that is interrupted only by an occasional convenience store. Near North Port, Warm Mineral Springs—a 2.5-acre lake well known for the therapeutic value of its 87-degree water—continues to be a popular tourist attraction. Next is the Port Charlotte area, a rapidly growing strip of stores, fast-food restaurants, and shopping centers. Here the Peace River empties into Charlotte Harbor; across the river is the town of Punta Gorda.

FIG. 5.5. The Shell Factory moved to the Tamiami Trail in 1951. Photo by the author.

Continuing south, development dissolves until after the Charlotte-Lee county line and arrival in North Fort Myers, where the setting is characterized by mile after mile of manufactured homes. The Shell Factory, at the same location since 1951, claims to be the "biggest outlet for sea stuff in the world" (fig. 5.5). The fiberglass conch shell beckoning tourists measures fourteen feet from tip to tip. At this point the trail veers to the east, where it crosses the Caloosahatchee River and arrives in downtown Fort Myers, where one is again surrounded by stores, restaurants, and offices.

FIG. 5.6. Everglades Wonder Gardens in Bonita Springs, virtually unaltered since its 1936 opening. Photo by the author.

The village of Estero was a major Lee County community at the turn of the century. In 1894 Dr. Cyrus Teed established the Koreshan Unity Settlement, a utopian community founded on the belief that the universe was a hollow sphere and man lived on the inside. His followers practiced celibacy, communal living, and communal ownership of property. A few miles south is Bonita Springs where the Tamiami Trail is still the main street through town. Bonita Springs is home to the Everglades Wonder Gardens, which is also a zoo and opened in 1938; it is one of the most intact roadside attractions along the route (fig. 5.6). Shortly after crossing the Lee-Collier county line, one arrives in the Naples metropolitan area.

In downtown Naples the trail turns ninety degrees to the east. Naples is home to Jungle Larry's Zoological Park and Caribbean Gardens, a fifty-two acre botanical and zoological preserve opened in 1954. The Trails End Motel foreshadows the transition in the terrain that lies ahead (fig. 5.7). From here development begins to fade, first to farmland, then to wetlands, and ultimately to the Everglades. Most of the land that the cross-state segment of the trail traverses is part of Collier Seminole State Park, Big Cypress National Preserve, or Everglades National Park. Signs of civilization are few and far between.

Thirty-two miles east of Naples is the tiny village of Ochopee, whose claim to fame is the nation's smallest post office, measuring seven feet by eight feet. About eighteen miles east of Ochopee is Monroe Station, one of the more notable structures along this stretch of the Tamiami Trail. Today, vacant and damaged by storms, Monroe Station is one of six original gasoline and rest stations constructed in 1928 (fig. 5.8).

Shortly after crossing the Dade County line, one is besieged with announcements for Indian villages and airboat rides (fig. 5.9). One can still witness alligator

FIG. 5.7. Trails End Motel, opened in 1946. Photo by the author.

FIG. 5.8. Monroe Station, one of six original Tamiami Trail gas station and rest stations, now located within Big Cypress National Preserve. Photo by the author.

FIG. 5.9. In the 1930s about a dozen "Indian Villages" lined the trail from Naples to Miami; this abandoned one is at Big Cypress Bend Boardwalk, eight miles east of Everglades City. Photo by the author.

wrestling at the Miccosukee Indian Village, as well as enjoy craft and cooking demonstrations, a museum, a restaurant, and a gift shop. Heading east past the Miami suburbs, the Tamiami Trail becomes Calle Ocho (Southwest Eighth Street), the heart of Little Havana, a neighborhood that is home to thousands of Cuban Americans. The Tamiami Trail ends at its intersection with U.S. 1 amid the office towers of Miami.

With the opening of Alligator Alley (State Route 84) in the late 1960s and the subsequent development of Interstate 75, the Tamiami Trail is no longer the only road connecting the east and west coasts of the lower peninsula of Florida.[24] The interstate system replaced the trail as southwest Florida's main thoroughfare, but it is not likely ever to equal the trail's contributions to the region's history or economy.

NOTES

This essay was reprinted from *SCA Journal* 15 (Fall 1997): 20–24. Used with permission.

1. Joe Hugh Reese, *History of the Tamiami Trail and a Brief Review of the Road Construction Movement in Florida* (Miami: Tamiami Trail Commissioners and Dade County Commissioners, 1928), 6–7.
2. "Trail a True Wonder," *Fort Myers News-Press,* April 13, 1955.
3. Doris Davis, "The Tamiami Trail—Muck, Mosquitos, and Motorists: A Photo Essay," *Tampa Bay History* 1 (Fall/Winter 1979):14.
4. Charlton W. Tebeau, *Florida's Last Frontier: The History of Collier County* (Coral Gables: U of Miami P, 1966): 221.

5. The association was active until October 4, 1920, the date of its last meeting. Reese, *History of the Tamiami Trail,* 7–8.
6. Reese, *History of the Tamiami Trail,* 8; and Davis, "The Tamiami Trail," 14.
7. Reese, *History of the Tamiami Trail,* 8–9.
8. Davis, "The Tamiami Trail,"15.
9. Reese, *History of the Tamiami Trail,* 3.
10. Davis, "The Tamiami Trail," 14.
11. Reese, *History of the Tamiami Trail,* 16
12. Federal Writers' Project, *Florida: A Guide to the Southernmost State* (New York: Oxford UP, 1939), 406; and Davis, "The Tamiami Trail,"15.
13. Davis, "The Tamiami Trail," 15.
14. Reese, *History of the Tamiami Trail,* 10–11.
15. Federal Writers' Project, *Florida,* 406.
16. Charlton W. Tebeau, *A History of Florida* (Coral Gables: U of Miami P, 1971), 380.
17. In 1925 the trail was designated State Highway No. 27. Reese, *History of the Tamiami Trail,* 17.
18. Federal Writers' Project, *Florida,* 74, and Tebeau, *Florida's Last Frontier,* 222.
19. Federal Writers' Project, *Florida,* 406.
20. Davis, "The Tamiami Trail," 16.
21. "Trail Flag Designed by Fort Myers Man," *Fort Myers Tropical News,* April 25, 1928.
22. "Motorcade, Monuments Mark Events," *Collier County News,* June 5, 1958; and "Tamiami Trail Celebration Set for June 7th," *Fort Myers News-Press,* May 25, 1958.
23. "A Trail of Progress," *Collier County News,* June 5, 1958.
24. The east-west segment of Interstate 75 was constructed in the mid-1970s and incorporated existing lanes from Alligator Alley (State Route 84) to carry eastbound traffic. Two westbound lanes were built on a vacant strip of land between State Route 84 and a canal.

CHAPTER 6

THE DOLLARWAY ROAD AND ROAD IMPROVEMENT DISTRICTS IN ARKANSAS, 1913–1921

CHRISTIE H. McLAREN

Prior to the development of state highway departments and the infusion of federal money in the late 1910s, local officials built roads. Dependent on residents for both money and labor, these roads varied greatly but tended to be of poor quality. Arkansas attempted an ambitious road building plan in 1913 to develop a concrete road system that initially met skepticism and opposition. Road names often were chosen to honor the dead: the Lincoln Highway, the Lee Highway, and the Jefferson Davis Highway, for example. Other names reflected regional or historical ties: the Egyptian Trail, the Old Spanish Trail, and the Dixie Highway, for example. However, the name "Dollarway" reflects the initial suspicions about costs to the rural southern communities that this substantial road would serve. One of many southern roads developed late in the Good Roads Movement just prior to the implementation of the state and U.S. routing systems, the Dollarway Road is an exemplary local road-building engineering achievement.

The coming of the automobile to Arkansas brought an awareness of the inadequacy of the state's rural roads. Although residents agreed that something should be done to improve the road situation, the question of how to create a superior road remained unanswered. While gravel and macadam paving had proven to be no match for the destructive force of mechanized travel, only time would tell which construction techniques would produce a lasting road surface. Concrete was growing in popularity as a road-building material in other regions, but in Arkansas its use was rare and limited to short sections of city streets. That is, until the successful completion of the state's first "Dollarway" road allowed auto travelers from near and far to enjoy a premium concrete surface.

Early road builders also contended with the issue of who would shoulder the financial burden for modernizing Arkansas's roads. Arkansas began using locally controlled road improvement districts to organize and finance construction. From the beginning, Arkansans recognized problems with the road improvement district system, yet it was not abolished. With no centralized authority these districts often overlapped one another, resulting in great tax burdens on property owners. Since the districts lacked centralized planning, completed roads did not always connect to each other.

Unfortunately, in Arkansas, the shortcomings of the road improvement district system were mistakenly placed upon the well-known road it built. Local tradition holds that the Dollarway Road was named because it was expensive for taxpayers, when it was actually constructed in the most inexpensive way then available.

THE DOLLARWAY ROAD

Travelers through Arkansas will find no highway markers designating the once-famous Dollarway Road (fig. 6.1). According to tradition, the Dollarway Road was so named because construction costs approximated one dollar per square yard or lineal foot of its nine-foot wide pavement.[1] Actual cost approximated $1.36 per linear foot.[2] Clues to its location are found only in the addresses of modern residents living along its path. Occasionally a narrow section of concrete can be found diverging from the path of its replacement road. In the early days, "sporting motorists drove long distances over primitive dirt and gravel roads or shipped their autos on flatcars to Pine Bluff for the thrill of spending a day or two traveling back and forth over 23 miles of bump, rut, and pothole-free 'super-road' at speeds up to 45 miles per hour."[3] Some farmers along the new road complained that the roaring automobiles frightened their livestock and that the concrete surface was too hard for their horses' hooves.[4] Others praised the road, saying, "it seems that the size of the load has ceased to depend on what the mules can haul, but rather depends on how much you can load on the wagon, and the ability of the mules to keep out of the way."[5]

Soon after the creation of Road Improvement District No. 4 by a special act of the Arkansas general assembly, bidding began for the construction of a macadam or concrete highway leading from the Pine Bluff city limits northwest to the Jeffer-

FIG. 6.1. An abandoned stretch of the Dollarway Road. The exact location and date of the photo are unavailable. Courtesy of the Arkansas State Highway and Transportation Department.

son County line. On July 31, 1913, commissioners of the road improvement district, chaired by Jefferson County judge C. M. Philpot, announced that Little Rock contractors Shelby and Bateman would construct the first "Dollarway Road" in the state. The road, as originally planned, would stretch twenty-two miles to the Jefferson County line, then a road of the same material would be constructed traversing six miles of Saline County to connect with a highway extending to Little Rock. According to the local newspaper, "the contract was awarded for Dollarway pavement, new in this section of the country, [which] calls for a five-inch concrete base, covered with bitumen and laid nine feet wide," to cost $160,000. Judge Philpot generously agreed to build the culverts and bridges out of county funds. The newspaper elaborates that "the cost to the individual property owners should not exceed $100,000," with the majority of the remaining cost to be funded by railroad taxes. Defending the choice of construction materials, the contractor is quoted as saying, "this is a new departure in road building in this part of the country . . . and should result in a permanent and lasting road." He ventured that this pavement would eventually replace macadam, which requires relaying every few years.[6]

Shelby and Bateman's explanation proved unsatisfactory to several Jefferson County taxpayers, however. On the following Saturday, August 2, they gathered at the courthouse to express their opposition to the use of Dollarway pavement. Through their attorney the property owners threatened to file an injunction to restrain the commissioners from beginning construction unless a cheaper material was chosen. After much discussion the meeting was adjourned until the following week, when a tax assessor would be available to show how costly the burden was likely to be for property owners along the road.[7] By the following Thursday the dispute was resolved. As stated in the local news, "One prominent land owner said that he did not believe the others would object to the use of the Dollarway pavement when he learned how small the assessments are."[8]

There were no more objections, and construction began on the new highway without delay in November 1913. The first few yards of concrete were laid at the end of the wood block pavement of West Pullen Street, near Bellwood Cemetery in Pine Bluff.[9] The completed route included four bridges, of "virtually indestructible" iron and concrete, at an estimated expense of five thousand dollars. Engineers offered a conservative estimate that two miles of road could be built each month, but with good weather they could advance at twice that rate.[10] True to their words, a two-mile-long section of the Dollarway opened to traffic in December.[11] In mid-March 1914, a delegation of civil engineers was treated to a luncheon along the progressing road.[12] By July enough of the road had been completed that a local newspaper reported several arrests were made for speeding along its superior surface.[13]

In the fall of 1914, a pioneer settler of Jefferson County brought a troubling fact to the notice of the commissioners. The final section of the proposed path of the road fell across the county line, outside the jurisdiction of the road improvement district. Investigation revealed that Jefferson County had continued maintaining the roadbed even after Grant County was created, encompassing the route in 1869. A solution was proposed that would include the Grant County land in question in the road improvement district. Although an account of the ensuing debate is unavailable, the path of the completed road lies on the Jefferson-Grant county line, following it until it intersects with the southwest corner of Pulaski County (fig. 6.2). Thus the finished road differed from its planned path. Rather than expanding west into Grant County and through the corner of Saline County, the road ran northwest, where it connected with the Pulaski County section of the Old Pine Bluff and Little Rock Wagon Road.[14] Despite the setbacks the concrete surface was completed on October 27, 1914, just eleven months after construction had begun.[15]

Once completed, the approximately twenty-three-mile-long stretch of concrete known as the Dollarway had the distinction of being the longest continuous length of concrete pavement in the nation.[16] It was the first use in Arkansas of reinforced concrete for road and bridge construction. A thin bituminous seal coat was applied to the concrete surface, marking its first use in the South.[17]

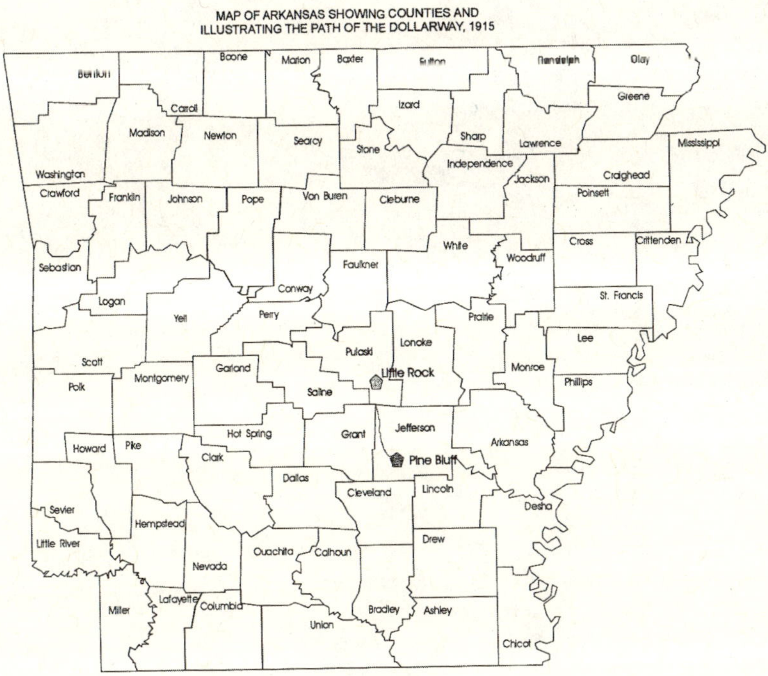

FIG. 6.2. Map of Arkansas showing counties and illustrating the path of the Dollarway, 1915. Courtesy of the author.

Technically the design for the Dollarway Road called for nine feet of five-inch concrete consisting of a mixture ratio of 1 part cement to 2.5 parts sand to 5 parts gravel. The edges of the concrete were constructed on a forty-five degree slope, giving a top width of 8.5 feet and bottom width of 9.5 feet. Eighteen inches of gravel and three feet of graded dirt were added on each side, constructing a total roadbed eighteen feet wide. The gravel shoulder enabled drivers to pass when they met another on the narrow roadway.[18] Two one-half-yard capacity stationary concrete mixers were the only pieces of machinery used on the job.[19]

Isaac Prather Shelby, engineer and contractor, was distressed by the poor horizontal alignment of the finished road in many places. He urged the road improvement district commissioners to permit him to correct the worst situations, such as right angle turns, even at his own expense. He was prohibited from doing this because state legislation specified that new pavement be placed only "on the road as now laid out."[20]

FIG. 6.3. The concrete surface of the Dollarway Road is still visible under the modern county road, named Reynolds Road. Courtesy of the Arkansas Historic Preservation Program.

Although modern State Highway 365 covers most of the original Dollarway Road, sections are still visible today (fig. 6.3). A roadside park featuring a short section of the original road was located south of Redfield, Arkansas, on State Highway 365. The park closed some years ago, but visitors can still see the path of the road winding through the trees. Jefferson County residents living near the park site use another section of the remaining road to access their property. Farther south on State Highway 365 is the best remaining section of the Dollarway Road still in public use. Along a 1.6-mile-long stretch, portions of the original roadbed can be glimpsed through the potholes of a dirt-covered local route. Now named Reynolds Road, this section also features two original reinforced concrete bridges (fig. 6.4).

FIG. 6.4. Remaining concrete bridges along the Dollarway Road. Courtesy of the Arkansas Historic Preservation Program.

Although not referred to as "Dollarway pavement," the construction process using a cement road base with a bituminous sealer was a popular method during the early twentieth century. Specifications for this type of pavement were outlined in the *American Highway Engineers Handbook* of 1919. New to the South, this technique had been perfected in other regions of the United States. The handbook claims Rhode Island was the first state to adopt bituminous concrete pavement as a standard type of construction for use on its highways, in 1909. It elaborated that "since 1910 there has been a rapidly growing appreciation of the inherent value of the many different types of bituminous concrete pavements for use on roads and streets."[21]

One reason for concrete's popularity was that it was relatively inexpensive when compared to alternative construction methods. The handbook lists the prices of bituminous concrete pavements in 1919 as ranging from $1.00 to $2.50 per square yard.[22] The average price for macadam, or broken stone, roads in various locations in the United States in 1915 ranged from $.57 to $1.40.[23] If the construction materials were available locally "the cost [for macadam construction] may be as low as $1.25 per ton or about $3,800 per mile of road 15 feet wide and 6 inches thick." However, "if the stone has to be shipped a long distance by rail and then hauled a considerable distance by team or truck, the cost of the stone rolled in place on the road may exceed $3.00 per ton or about $10,000 per mile.[24] Experts recommended that macadam roads be resurfaced as soon as the top layer of stone was worn through, adding to the expense.[25] According to an article written by one of the Dollarway's engineers, "Jefferson County [had] absolutely no natural road materials within its boundaries."[26] Therefore a macadam road would have been more costly to construct than a permanent concrete surface.

The designers of the Arkansas Dollarway Road recognized the value of concrete road construction. Experiments with that method were not unknown in Arkansas. The pavement technique had been utilized on sections of city streets in Pine Bluff and Little Rock. According to the local paper, "West Tenth street in Little Rock is paved with this material and East Eighteenth avenue in Pine Bluff, from Main to Georgia streets is paved with Dollarway pavement, as is Oak street from Sixth to Eighth avenues."[27] The builders knew that the route would be subjected to a continuous flow of through traffic between Little Rock and Pine Bluff in addition to local use. Road-building materials such as macadam and gravel had failed under continuous auto traffic.[28] They sought to create a lasting road surface and chose "Dollarway pavement" for its ability to survive heavy use.

According to the 1919 *American Highway Engineers Handbook,* there were several popular methods for constructing concrete roads. Likewise there were numerous specifications for the mixture of concrete, sand, or asphalt that the road surfaces comprised. Although records of the exact process utilized to seal the concrete on the Arkansas Dollarway have not been located, available sources state that the completed concrete was covered with a thin coating of bitumen. A similar process is described in the handbook. A hot liquid is spread over finished concrete, then sand is rolled onto the liquid to create a waterproof, nonskid surface.[29] Durability aside, the selling point of Dollarway pavement, as far as the taxpayers were concerned, was that it was inexpensive. The technique created a permanent, superior surface for an agreeable price.

THE LEGACY OF THE ROAD IMPROVEMENT DISTRICT SYSTEM

The search for an effective construction material for road surfaces predates the widespread use of automobiles in Arkansas. As in other parts of the nation, Arkansas's Good Roads movement was founded by bicyclists. Poor road conditions were an issue for bicyclists who struggled over ruts and potholes and became mired in mud. Their efforts led to the founding of the Good Roads League of the State of Arkansas in 1896. Soon auto enthusiasts dominated this organization. In response to public support, the state legislature began looking into the question of providing adequate funding to construct better roads. The only legislation in place concerning highway maintenance at this time required healthy men between twenty-one and forty-five years of age to contribute five days per year of road work or the monetary equivalent.[30] Little was accomplished until 1907, when the Arkansas legislature passed an act allowing the formation of local road improvement districts.[31] These poorly supervised districts with no centralized construction plan would soon prove problematic. Eventually the road improvement district system would be blamed for widespread indebtedness, finally resulting in scandal.

After 1907, by legislative act, counties were authorized to establish road improvement districts at the request of a majority of landowners along a local route. Taxes were accessed against property lying adjacent to the road to fund construction. Various legal questions remained obstacles, however, meaning few road improvement districts were created under this early law.[32] Notably Pulaski County, home of the state capital, and Jefferson County, of Dollarway Road fame, were the locations of several of the first few districts created.[33] When the Arkansas Highway Commission was founded in 1913, the shortcomings of the road improvement district system were already apparent. The commission acknowledged several problems in its first biennial report, including poorly trained overseers and a lack of centralized highway planning that resulted in little improvement at a great expense. "Under our present lack of system, an enormous sum of one million, two hundred thousand dollars per year is being spent from the general three-mill tax alone, with no great improvements over conditions of the past."[34] By February 1915 an audit committee reported misuse and waste within the Arkansas Highway Department. The audit committee accused the department of granting free registration tags to officials, depriving the state and county of revenue and abusing its power. The committee also charged that the department used state funds to make unauthorized purchases of autos, gasoline, and other auto-related expenses. The department would make no further response on the charges, except that "it illustrates the fact that automobiles are expensive."[35]

Also in 1915, the legislature tried again to pass an effective road law. The Alexander Road Law allowed "neighbors along a particular route [to] form a road improvement district, contract out for the construction, borrow the money to finance the project, and in effect, mortgage their property as a debt guarantee."[36] No provision was made to ensure a unified plan for construction throughout the state, resulting in widely varying road conditions from county to county. The Alexander Road Law, coupled with the Federal Aid Road Act of 1916 aimed at improving rural post roads, paved the way for the creation of hundreds of road improvement districts.[37] By 1921 more than five hundred districts had been created. "State and federal aid was available for about ten percent of the highway construction costs, the remainder to be met by the real property tax authorized by the Alexander law."[38] The new Federal Aid Act of 1921 tried to deal with some of the problems Arkansas and other states had experienced under the 1916 law. The 1921 act provided for the construction of a national road system consisting of interconnected state highways. To facilitate this task, it required that a state highway department supervise all construction and maintenance.

Meanwhile the scandalous legacy of the road improvement districts in Arkansas was attracting national attention. The *New York Times* reported that "land owners in numerous instances [were] being taxed amounts that equal or even exceed the total gross incomes of their property."[39] Compounding the problem, districts were allowed to overlap, making property owners liable for taxes in more than one district. A grand jury in Little Rock reportedly charged that "gross extravagance and waste has prevailed" in Arkansas under the use of road improvement districts.[40] Debated in the U.S. Congress, the issue resulted in a mandate that federal funds be withheld from states without a centralized means of supervising and planning road construction.[41]

Yet, despite the efforts of a road champion, Gov. Thomas Chipman McRae, the Arkansas legislature was unable to enact the required laws. Reportedly the legislature had become involved in controversies over the allocation of automobile taxes and local control of road construction. Members of the County Judges Association, who wished to maintain local control of their roads, successfully blocked the passing of an effective bill.[42] Inevitably, in 1923 federal officials recalled technical personnel and withdrew construction funds from the state. In response Governor McRae called a special session of the general assembly in September 1923 and demanded that it pass the necessary legislation to bring about an end to the crisis.[43] The result was the Harrelson Road Law, which eased the burden of property owners somewhat by increasing user taxes and granted supervisory status to the reconstituted State Highway Commission.[44] Arkansas now possessed the prerequisites necessary to develop a state system of roads adequate to become part of a national highway system. Property owners, however, would have to wait until the Martineau

Road Law of 1927 to be relieved of the responsibility for funding road construction. With this act the legislature placed the financial burden on the road user rather than the landowner.

In Arkansas, as in other parts of the nation, the first roads built were themselves experiments in road construction. Likewise, the systems used to finance and organize road construction were untested. The Dollarway Road remains as an example of an instance when the early use of the road improvement district system functioned. With the approval of local taxpayers, the road improvement district commissioners and their engineers created a durable road to facilitate travel between their community and the state's capital. Sections of the original road surface remain as testimony to the ability of this pavement type to weather auto traffic.

Ultimately the road improvement district system failed to be an equitable solution to the Arkansas road-building problem. Use of this system had resulted in ruinous indebtedness for rural Arkansans and finally had embroiled the state in scandal. Arkansas remained dedicated to the road improvement district system, however. Not even the national attention resulting from the financial situation created by the road improvement districts brought about change. Local officials were the foundation of the problem, with county judges unwilling to relinquish control even after federal support was repealed for Arkansas road projects do to the state's lack of a centralized highway program. But the dark days resulting from the road improvement district system's deficiency could not erase the Dollarway. It remains an engineering accomplishment of which Arkansans can be proud.

NOTES

1. Arkansas State Highway and Transportation Department, *Historical Review: Arkansas State Highway Commission and Arkansas State Highway and Transportation Department, 1913–1992* (Little Rock, AR, 1992), 15.
2. Hugh H. Humphries, "County Highway Engineer Tells of Progress in Jefferson County," *Pine Bluff Commercial,* April 24, 1915.
3. Arkansas State Highway and Transportation Department, *The Dollar Way* (Little Rock, AR, n.d.), 1–3.
4. James W. Leslie, *Pine Bluff and Jefferson County: A Pictorial History* (Virginia Beach, VA: Donning County Publishers, 1997), 142.
5. Humphries, "County Highway Engineer," *Pine Bluff Commercial,* April 24, 1915.
6. "First Dollarway Road in State," *Pine Bluff Commercial,* July 31, 1913.
7. "Opposition to Dollarway Road," *Pine Bluff Commercial,* August 2, 1913.
8. "Opposition to Dollarway Less," *Pine Bluff Commercial,* August 7, 1913.
9. "Will Soon Open Dollarway Road," *Pine Bluff Commercial,* December 20, 1913.
10. "Begin Work on New Highway," *Pine Bluff Commercial,* November 6, 1913.
11. "Will Soon Open," *Pine Bluff Commercial,* December 20, 1913.
12. "Arkansas Civil Engineers Here," *Pine Bluff Commercial,* March 14, 1914.

13. "Several Arrests for Speeding on Dollarway Road," *Pine Bluff Commercial,* July 29, 1914.
14. Arkansas State Highway and Transportation Department, *The Dollar Way,* 1–3.
15. "Dollarway Road in Grant County" and "Complete Concrete for Dollarway," *Pine Bluff Commercial,* October 3 and 27, 1914.
16. Arkansas State Highway and Transportation Department, *The Dollar Way,* 1–3.
17. Ibid.
18. Leslie, *Pine Bluff and Jefferson County,* 142.
19. Ed Baxter, "Dollarway Road" (National Register of Historic Places Nomination Form, Jefferson Co., Near Redfield), Arkansas Historic Preservation Program Files, 1973, Little Rock.
20. Arkansas State Highway and Transportation Department, *The Dollar Way,* 1–3.
21. Arthur H. Blanchard, ed., *American Highway Engineers Handbook* (New York: John Wiley and Sons, 1919), 849.
22. Ibid., 919.
23. Ibid., 589.
24. Ibid., 586.
25. Ibid., 592.
26. Ibid., 919.
27. "First Dollarway," *Pine Bluff Commercial,* July 31, 1913.
28. "Humphries," *Pine Bluff Commercial,* April 24, 1915.
29. Blanchard, *American Highway Engineers Handbook,* 853.
30. Arkansas State Highway and Transportation Department, *Historical Review,* 5–15.
31. Jerry E. Hinshaw, *Call the Roll: The First One Hundred Fifty Years of the Arkansas Legislature* (Little Rock, AR: Rose Publishing, 1986), 82.
32. Arkansas State Highway and Transportation Department, *Historical Review,* 14.
33. John Hume, "The Auto Age in Arkansas, Part II," *Arkansas Highways,* Summer 1977, 13.
34. State of Arkansas, Department of State Lands, Highways and Improvements, *First Biennial Report* (Little Rock, AR, 1914), 16.
35. "Highway Department Is Under Fire," *Pine Bluff Commercial,* February 26, 1915.
36. Arkansas State Highway and Transportation Department, *Historical Review,* 24.
37. William B. Mayes Jr., "Arkansas Highway Development, 1907–1927," master's thesis, State College of Arkansas, 1968, 15.
38. Ibid.
39. "Arkansas Totters under Road Taxes and Czar-like Rule," *New York Times,* March 26, 1921.
40. "Widespread Alarm in Arkansas on Tax for Road Projects," *New York Times,* March 27, 1921.
41. Hinshaw, *Call the Roll,* 82.
42. Mayes, "Arkansas Highway Development," 19.
43. W. David Baird, "Thomas Chipman McRae," in *The Governors of Arkansas: Essays in Political Biography,* ed. Timothy P. Donovan and Willard B. Gatewood (Fayetteville: U of Arkansas P, 1981), 156–57.
44. Ibid., 158.

CHAPTER 7

THE DAUGHTERS OF THE AMERICAN REVOLUTION, ROANE F. BYRNES, AND THE BIRTH OF THE NATCHEZ TRACE PARKWAY

SARA AMY LEACH

The Natchez Trace Parkway was established and built as a commemorative property in the South. Largely through the efforts of women's organizations and one woman in particular—Roane F. Byrnes—the National Park Service undertook this planned development through three states. While economic development in communities along the Dixie Highway was a significant component in establishing that route, early proponents of the Natchez Trace Parkway were only interested in getting tourists and economic development to Natchez, Mississippi. Yet, as with the Dixie Highway, the establishment of the parkway was not an easy task, and there were conflicting ideas as to how it should be accomplished. Ultimately the Natchez Trace Parkway became an example of commemorating early southern history, scenic automobile travel and tourism, and the partnership of private individuals with the federal government.

The Natchez Trace, or Old Trace, became the oldest federally funded road in the country in 1801, when President Thomas Jefferson assigned the money and military to improve it, initially as a post road into the Old Southwest. The trace was in use from pre-Columbian times through the peak years of 1780–1820. Its demise came in 1830, when the invention of the steam engine offered boatmen and other travelers a faster and safer passage from Natchez, Mississippi, northeast to what would become Nashville, Tennessee. The road—really a series of foot trails—was about five hundred miles long. Its travelers included Native Americans, post riders, missionaries, enslaved Africans, military troops, Andrew Jackson, and Meriwether Lewis. Between 1830 and the early twentieth century, much of the Natchez Trace fell into disuse, while about 60 percent became part of local transportation systems.

The Natchez Trace Parkway was established in 1938 in order to commemorate the Natchez Trace. It is one of nine parkways in the National Park Service (NPS).[1] The Natchez Trace Parkway today is 445 miles long and averages 800 feet wide and passes through Mississippi, Alabama, and Tennessee (fig. 7.1). A two-lane motor road meanders independent of the Old Trace, portions of which have been preserved and maintained for trail and interpretive use. The federal parkway contains about ninety-seven miles of the Old Trace.[2] Where the old and the new roadways intersect, visitors are able to leave their vehicles and walk over the ancient route.

The ancient Natchez Trace came to be celebrated in a modern parkway, after the patriotic women of Mississippi—first the Daughters of the American Revolution (DAR), then Roane Fleming Byrnes as head of the Natchez Trace Association—conceived and nurtured the project until it took on a life of its own. Rooted in the spirit

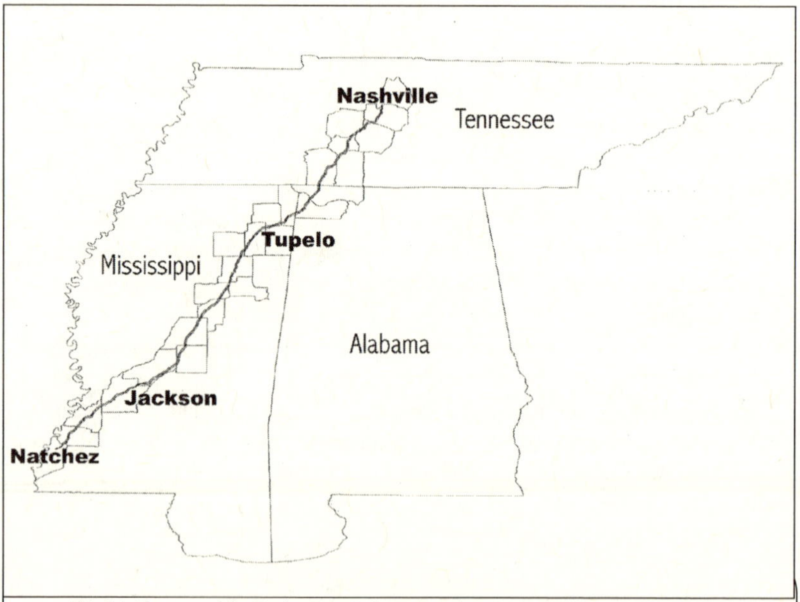

FIG. 7.1. The route of the Natchez Trace Parkway through Mississippi, Alabama, and Tennessee. Map by the author.

of Ann Pamela Cunningham and the Ladies of Mount Vernon, the Old Trace preservation movement was perfectly timed to encompass an outgrowth of romantic nationalism, private local initiatives, and cultural marketing that were newly successful in the city of Natchez. The movement was also able to take advantage of federally funded initiatives.

MISSISSIPPI SOCIETY OF THE DAUGHTERS OF THE AMERICAN REVOLUTION

The Mississippi State Society of the Daughters of the American Revolution (DAR) was established in 1896, when a dozen ladies organized the first chapter in Natchez. Mrs. Egbert Jones, the first DAR member in Mississippi and state regent from 1906 through 1908, suggested the DAR mark the Old Natchez Trace at the organization's first statewide conference nine years later. The plan called for a granite marker to be placed in each county through which the Old Trace passed. Jones's idea was adopted in 1908, and the first granite marker was set that year in Tishomingo County; it was followed by the Natchez marker in 1909 and a joint Alabama-Mississippi marker at the state line, placed in 1913. A total of seventeen stone boulders in Mississippi and Tennessee were erected by 1933.[3]

The DAR's influence prior to the creation of the NPS unit should not be overlooked. In 1934 the Department of the Interior authorized the Natchez Trace Parkway Survey. This study, published in 1941, remains the earliest and most accurate account of the route's history. Part of the 1943 survey legislation states that the Natchez Trace "is one of the nation's most famous old roads, and has been marked by handsome boulders with suitable inscriptions by the DAR at great expense."[4] It should be noted that between 1911 and 1918 the Texas DAR undertook a similar marking project along El Camino Real from Louisiana to Mexico, which may have been inspired by the Mississippi marking project.[5]

Although 1934 marked the beginning of officially documented post-DAR activities to commemorate the Old Trace, at least one other effort had already occurred. In 1915 the Natchez Chamber of Commerce initiated a movement to pave the Old Trace, then advocated as the means of preserving the route. Although there is limited documentation to indicate what actually occurred, it is known that a convention was held January 19, 1916, in Kosciusko, Mississippi. The first incarnation of the Natchez Trace Association may have been organized here, with the DAR assigned two permanent seats on its board of directors.[6] As the DAR marking project concluded in the early 1930s concurrent to the germination of the Natchez Trace Association, relations between the two groups and the accomplishments of individuals in both camps are unclear. Subsequently direct DAR support of the parkway funding and an article about the Natchez Trace appeared in *DAR Magazine* in February 1964.[7]

NATCHEZ TRACE ASSOCIATION

By most accounts, the Natchez Trace Military Association (NTA) formally organized on January 21, 1934, in Jackson, Mississippi. It was a self-described "civic and patriotic organization . . . trying to induce the Federal Government to pave the ancient Trace and preserve it as a National Parkway."[8] Within a few months the group officially dropped "military" from its name because the Department of the Interior was to manage the project rather than the War Department, which supervised highways built purely for defense purposes.[9] Each county through which the Old Trace ran organized a chapter of the NTA, elected officers, and collected dues that went toward expenses—delegate travel to Washington, D.C., supplies, and the many telegrams sent to legislators, government officials and powerful constituents. Roane Fleming Byrnes began as corresponding secretary of the Adams County chapter, but she was elected vice president of the state organization in July 1934, and president in March 1935—a position she would hold for thirty-five years (fig. 7.2).

FIG. 7.2. Roane Fleming Byrnes standing on a section of the Natchez Trace Parkway. Courtesy of Natchez Trace Parkway, National Park Service, Tupelo, MS, Interpretive Photo Collection, photo no. 5497.

That the DAR conceived the idea to save the Natchez Trace was never in question, but there is little documentation that details the transition between the women of the DAR and NTA membership. They did overlap, but perhaps less than amicably. In 1934 Lucille Mayfield, president of the new NTA and a DAR representative, lived in Washington, D.C., where she had firsthand contact with Capitol Hill. She asked Byrnes if she could invite Jones to attend an upcoming regional NTA meeting because, "This is her due as she had the vision, 30 years ago, to save the Natchez Trace for Mississippi. It was the first project for the DAR to take up, and I feel she would appreciate the recognition now."[10] The response to Mayfield infers the tension that must have existed between the two groups, as the role of the NTA eclipsed the DAR. Jones was one of "two members in DAR whom I think will not do a great deal to help us," cautioned one association official. "Neither of them appreciated being elected honorary vice-president and they told me so in no uncertain words."[11]

While the idea of paving the Old Trace into a modern highway was popularly expounded, the project had to first be given legal authority. Again multiple parties claim responsibility. The credit for approaching Congressman Jeff Busby of Mississippi's Fourth District may belong to amateur journalist "Colonel" Jim Walton. According to Walton, the idea for the modern Natchez Trace Parkway came at a November 1932 U.S. Good Roads Convention in Beaumont, Texas. Walton shared the idea with participants there and others in Mississippi before purportedly contacting Busby. Whether it was political posturing or underestimating the power of the ladies behind it, on asking Busby to introduce a bill into the House of Representatives to pave the Old Trace, Walton is said to have added that while "he was certain such a bill would not pass . . . he thought the women who had worked hard on the Natchez Trace project deserved the compliment."[12] Busby introduced two bills in 1934: one for a fifty-thousand-dollar survey of the Natchez Trace, a second for 25 million dollars in construction funds. Only the survey bill passed, and notably against the wishes of Interior Secretary Harold Ickes. Association President Mayfield credited Sen. Pat Harrison of Mississippi with keeping the bill afloat and getting it passed. Harrison, who was the chair of the powerful Senate Finance Committee, and other members of the Mississippi delegation paid a convincing visit to President Roosevelt, who signed the legislation May 21, 1934.[13] Now that they had the politics, planning, and funding assured for the next year or so, the NTA began a full-scale promotional campaign through special events, media, and active lobbying of the Mississippi delegation that continued throughout the decade.

Approval of the survey project resulted in a major public event in July 1934, when Natchez and Adams County, Mississippi, hosted a grand rally "for the purpose of enlisting interest in and support for paving the Natchez Trace. As you know

. . . this is the first link of the proposed great international highway of strategic value connecting South and Central America with this country, and following the direct route of historic interest from the southwest to Washington and the east."[14] More than 150 delegates from Mississippi, Tennessee, Louisiana, and Texas attended the rally, as well as politicians from these states. Activities included a luncheon for more than 220 persons, a conference, an auto tour, an afternoon tea at Auburn House, and speeches (including an address by Busby), all capped by a 10 P.M. ball at the Eola Hotel. Much of the day's events were broadcast live on the radio.[15]

FIG. 7.3. Cartoon, April 21, 1935. Natchez Trace Parkway Collection. Permission granted by The Times-Picayune Publishing Corporation. All rights reserved.

FIG. 7.4. Byrnes on Old Colony Road, a section of the Natchez Trace Parkway, about three miles south of Port Gibson. Courtesy of Natchez Trace Parkway, National Park Service, Tupelo, MS, Interpretive Photo Collection, photo no. 5499.

In an unusual move, although the parkway had yet to be established, funds to begin construction were disbursed through the Emergency Relief Appropriation Act of 1935, and construction began in 1937. Three discontiguous sections of Mississippi parkway were started that fall, again kicked off with pomp and ceremony, especially the ground breaking at Madisonville, where Governor White shoveled the first bit of dirt into a silver chalice held high by Byrnes. Roadwork progressed, and on May 18, 1938, the parkway became an official unit of the National Park Service. First-year funding was set at 1.5 million dollars. President Roosevelt sent Byrnes the pen he used to sign the bill creating the parkway, and it became one of her most precious mementos.[16]

Two proposed extensions of the project were short-lived. Although the Old Trace would link Nashville to Natchez according to the legislation, the survey report mentions the "possibility of extending this project north to Louisville as a national parkway" without historic justification. A more likely addendum was to the west. Byrnes and others quickly embraced the idea of extending the parkway through Louisiana and Texas, to terminate in Mexico City. The idea began as a cultural proposition mingled with the Pan-American Highway, but as World War II approached, justification angled toward "a speedway in national defense" and protection of the Panama Canal. Even further ahead, because Oak Ridge, Tennessee, was nearby, in the 1950s the parkway was dubbed an "atomic highway."[17] Abundant press coverage during these years harnessed the history and adventure of the Natchez Trace as well as propositions to take it as far as it could go, although the latter never completely coalesced on the ground.

Texas and Louisiana Natchez Trace associations, both anxious to lure tourist dollars to their states, never agreed on an exact route to link Natchez and Mexico. The absence of details did not faze Byrnes, however, and the National Park Service did not address the concept for another decade. To fuel the idea, Byrnes traveled to Mexico on the "Know Mississippi Better Train," and with a special courtesy bestowed by the U.S. embassy, she personally invited the Mexican president to Natchez to inaugurate work on the parkway. Although he could not attend, in October 1937 the inauguration of the parkway featured a Mexican delegation in Natchez that was reported as the "first international celebration [in Natchez] in 250 years."[18] On arrival General Mugica was given a key to the city, and the party was escorted through town by a sixty-piece band to Byrne's Ravennaside home for a reception. The next day businesses and schools closed. A parade passed through the downtown area and up to Brandon Hall plantation at the southern terminus of the trace, where "thousands" of visitors enjoyed a barbecue. The first trees, representative of the participating states and Mexico, were planted and labeled here.

Bills were drafted as early as 1939 to fund a study of a parkway extension across Louisiana and Texas and north to Louisville, but they were finally rejected by the interior secretary in 1949 because the original 1934 survey legislation stopped at Natchez, and the "historical associations or scenic possibilities" of any extension were lacking.[19]

Byrnes and key NTA leaders kept the organization connected and proactive. The NTA met annually, and the gatherings were legendary. "In her customary floppy hat and unabashed optimism, Mrs. Byrnes steered the meeting through two and one-half hours of unrehearsed decorum, the likes of which Roberts never had in mind when he wrote his rules of order," wrote the journalist W. F. Minor, who covered the group for more than two decades. "The rules were not Roberts, they were Mrs. Byrnes, and nobody expected anything different. . . . [She was] an irrepressible optimist [who] badgered congressmen, senators, the National Park Ser-

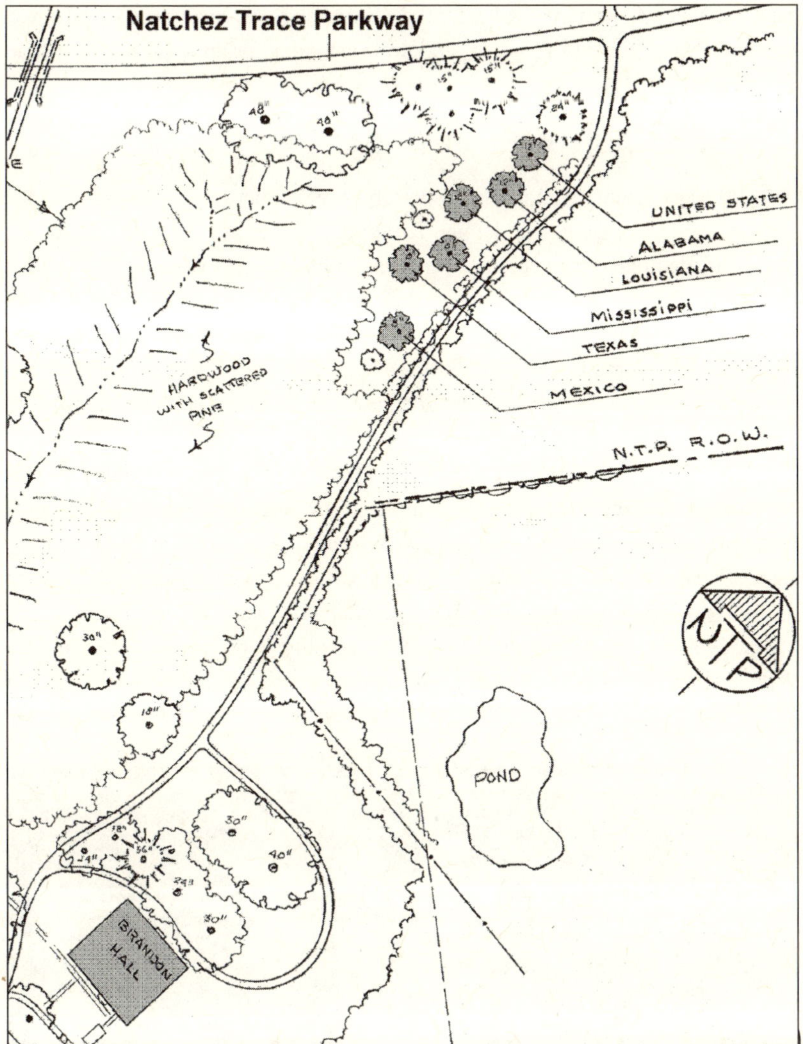

FIG. 7.5. Detail of site plan showing commemorative trees planted at southern terminus of the future Natchez Trace Parkway in 1934. Natchez Trace Collection, modified by the author.

vice—anyone—to carry out the dream of the rebirth of the old Natchez Trace for modern day travel."[20]

"It was through the earnest and intense efforts of Mrs. Byrnes that the Natchez Trace Parkway became a reality," lauded the *Natchez Democrat*. The paper ranked it her number one accomplishment in her obituary, which began, "Mrs. Roane Fleming Byrnes, president of the Natchez Trace Association in Mississippi, outstanding civic leader of Natchez; untiring worker for the cultural and economic growth and development of Natchez and this area, and an outstanding example of true Christian womanhood."[21]

"SWEET AUNTIE": MARY ROANE FLEMING BYRNES

In 1890 Mary Roane Fleming Byrnes was born a southern aristocrat, as part of a multigenerational Natchesean family. The timing of her arrival at the end of the tradition-laden nineteenth century, when a woman's success was measured by her marriage and social accomplishments, was the foundation for an atypical blend of abilities. An acknowledged "belle" in her youth, Byrnes was an accomplished writer and a popular storyteller known fondly as "Sweet Auntie." Her involvement with patriotic and community projects included the new Natchez Garden Club in 1929, of which she was a charter member.

In 1931 the Mississippi Federation of Garden Clubs was scheduled to visit Natchez, but a freeze threatened to doom the event. Rather than cancel it, twenty of Natchez's renowned antebellum homes were opened to visitors. The tour's meteoric success coalesced into the Spring Pilgrimage, which still flourishes today, inviting tourists to "step into the past with Natchez where the Old South still lives."[22]

Houses and history proved a salable commodity in the wake of the Colonial Williamsburg model and early historic-district designations underway in Charleston and New Orleans. Byrnes saw a future for Natchez's enclave of antebel-

FIG. 7.6. Byrnes family home, Ravennaside, in the city of Natchez. Photo 1998 by the author.

FIG. 7.7. Byrnes (second from left), stands in front of an interpretive display at the Ridgeland, Mississippi, museum with Malcolm Gardner (left), the parkway superintendent, and two others. Courtesy of Natchez Trace Parkway, National Park Service, Tupelo, MS, Interpretive Photo Collection photo no. 9262.

lum mansions, and she urged the Natchez Garden Club to purchase Connelly's Tavern/House on Ellicott Hill, built circa 1790, and restore it. This was one of the city's most significant structures, but its historic role for eighteenth-century travelers also served Byrne's real passion. She justified its restoration "for many visitors who would soon be traveling down the Natchez Trace."[23] One space in the building was, for a time, designated the Natchez Trace Room and contained memorabilia, including the silver chalice and groundbreaking dirt. Despite obvious allegiances it was not until later in Byrnes's life—after the connections were eclipsed by real deeds—that she formally affiliated with the Colonial Dames of America (mid-1940s), United Daughters of the Confederacy (1958), DAR (1961), and the Order of the First Families of Mississippi, 1699–1817 (1970).[24]

Ravennaside was Byrnes's lifelong residence (fig. 7.6), inherited from her parents and shared with her acquiescent husband, Ferriday. A sprawling circa-1910 Colonial Revival with ornate, high-ceilinged rooms, the residence was perfect for the frequent entertaining of society, political, and National Park Service figures. During the Depression years a writer friend joked that the Natchez Trace was paved with Byrnes's veal loaf. Byrnes herself recalled that "meatloaf and moonshine" were the best almost anyone could afford in those days.[25] The mansion contained two spatial testimonies to Byrnes's devotion to the parkway project. Mounted on a wall of the narrow pantry dubbed "the War Room" was a schematic site plan of the parkway. As sections were constructed, she inked in the paper alignment. Another space, the Trace Room, was dominated by a decorative photographic mural measuring about five feet tall that was mounted below the cornice. Installed during the 1950s and removed in 1997, the images by Mabel Lane were enlarged and mounted on canvas in Chicago, then hand-colored using oil paint.[26]

THE NATIONAL PARK SERVICE DESIGNS A NATIONAL PARKWAY

At a time when the American historic preservation movement was turning the corner from avocation to profession, the amateur strides made by the women of Natchez in the 1930s and the city's relationship to the Old Trace may have influenced NPS officials involved in parkway development. In a 1934 planning meeting, NPS Chief Historian Verne E. Chatelaine described the city as a "gold mine" and recommended "some method of conserving and preserving and developing in a historical way the plans for the City of Natchez. I am interested in the development of that more than any other thing," he confessed.[27] Byrnes recognized the same potential. In a 1937 speech she said, "Tourists will pour over [the parkway] by the tens of thousands, just as they do in Yellowstone and all other great National Parks. And Natchez is to be the southern terminus of this road, which bears its name. There is no way to compute what this will mean to us—Natchez will be more famous than she ever was—even in antebellum days."[28]

While Byrnes and the NTA were lobbying the public and politicians alike, in Washington, D.C., park officials struggled with more fundamental aspects of the parkway. The original Natchez Trace was researched and mapped by the NPS Branch of Historic Sites and Buildings, and the Branch of Plans and Designs, with the cooperation of John Nolen, a noted planner who served as consultant to the NPS. The Bureau of Public Roads flagged the alignment in the field. The states acquired land and deeded it to the federal government. Planners now had to decide whether to attempt a route "with only minor deviations, [which] may follow faithfully the line of the old road for merely historical reasons; or second, the parkway itself should . . . [follow] the general directions of that historic route, but essentially satisfying the technical and esthetic standards of modern parkway construction."

The summary report of the survey was submitted in June 1935, and its recommendations recognized the qualities and purpose of the parkway. Although the scenery was generally classed as "normal," the southernmost portion "would introduce the traveling public to an area where scenery, architecture, and historical associations combine to portray graphically the story of the Old South . . . the scenic beauty, historic associations of the Trace, and the existing need for a good highway from Jackson, the State capital, to Natchez, make the construction of this parkway justifiable."[29]

Even as the government was studying the trace, talks ensued about its historical merit compared to other roads, and the precedent its treatment could establish for the Oregon Trail or Cumberland Gap. Its existing condition, 61 percent in active use, was the dilemma.[30] Did current future traffic estimates justify a new highway in a state that lagged behind in road construction, or did the historical merit warrant a recreational parkway? Another variable was the submarginal lands program, whose funds could be appropriated to acquire acreage, at least in Mississippi. By the time the master plan was approved about 1940, however, the commemorative role and broad design concepts that promoted an uninterrupted, scenic, and noncommercial motoring experience were in place.

CHANGING OF THE GUARD

Parkway construction continued apace between 1937 and 1941, then from the end of World War II through the late 1960s and the NPS's emphatic Mission 66 program, which saw much of the road finished. By this time, however, the concept of recreational parkway construction was threatened. The Vietnam conflict precluded serious funding, and the interstate system was the first priority of the former Bureau of Public Roads, now the Federal Highway Administration. Parkway Superintendent Malcolm Gardner, an assertive historian who had led the trace since 1936, retired in 1967. With several hundred miles of motor road and some facilities complete, the NTA's mission and membership was adrift.

Byrnes's tenure surpassed even that of Gardner. She appeared at benchmarks of progress from the groundbreakings of the 1930s and 1940s, to road and facility dedications in the 1950s and 1960s. When she announced at the 1967 NTA meeting, "I want to ride on the Natchez Trace all the way before I have to ride on the golden street," the new parkway superintendent, Joseph Rumburg Jr., responded.[31] The following May, an NPS entourage took the aging Byrnes and a cousin on a four-day tour of more than 300 miles of parkway that had been completed in little more than thirty years. "Never have I had a more wonderful and delightful time," she later wrote, "riding over miles and miles of completed parkway."[32] Interestingly, given her astounding influence on a motor road, Byrnes never learned to drive.[33]

Two years later, in 1970, the eighty-year-old Byrnes's health began to fail. She resigned as NTA president and died just days before a tribute to her was scheduled by the organization she had led for thirty-five years. Happily, during Byrne's lifetime she was recognized for her work by local organizations, and in 1966 she received the American Association for State and Local History's Award of Merit. With Byrnes's passing the NTA immediately reorganized with the goal of "reactivating . . . and strengthening . . . its efforts to obtain completion of the Trace."[34] Its headquarters moved from Natchez to near park headquarters in Tupelo, Mississippi, and active construction, and NTA county chapters were disbanded. Thereafter came a brief spurt of energy and pronouncements to "Finish the Trace," but times had changed. Even more dormant was the relationship between the DAR and the NPS.

EPILOGUE

In 1905 the Mississippi DAR conceived of the idea to mark the Old Trace as a patriotic gesture that coincided with a national effort to improve U.S. roads and build recreational parkways. The DAR continued to take a role in efforts of the early 1930s, which coalesced in the NTA organization. Led by the indomitable Roane Byrnes, this group spent its first three decades ardently lobbying, promoting, and justifying to the public, Congress, and federal officials why legislation and funding for a national parkway to commemorate the Old Trace was essential. Seeing the parkway to fruition was Byrnes's singular goal. But that was not the end of the story.

In 1998 the Natchez Trace Parkway celebrated the sixtieth anniversary of the authorization of the NPS unit. A May celebration included members of the renamed Natchez Trace Parkway Association and DAR officials. This event reinvigorated the drive to complete the parkway, which was stalled in two Mississippi locations, around Jackson and at the southernmost terminus leading into Natchez. As with many a government project, inadequate funding and politics were the obstacle. However, on May 21, 2005, the Natchez Trace Parkway was finally completed. On a steamy day more than a thousand people—including Federal Highway Administration and NPS officials, legislators, and state and local officials—gathered in Natchez for speeches, the unveiling of a DAR plaque, and a parade of automobiles that ended at Melrose Plantation. The activities evoked the heady festivity in Natchez when the original survey was approved in 1934. The credit for this last resurgence goes to, again, the DAR and the Natchez Trace Parkway Association, whose membership had unyieldingly called to "finish the parkway." It is a story of an old road and a new road coming to an end at last.

NOTES

1. It is the second longest after the Blue Ridge Parkway; others are the George Washington Memorial Parkway, Clara Barton Parkway, Rock Creek and Potomac Parkway, Baltimore-Washington Parkway, and Suitland Parkway in the Greater Washington, D.C., area and the John D. Rockefeller Jr. Memorial Parkway and Colonial Parkway.
2. According to List of Classified Structures data, September 1999, Natchez Trace Parkway (hereafter, NATR).
3. *History of the Mississippi State Society, Daughters of the American Revolution, 1896–1996* ([Kosciusko, MS]: The Society, 1996), 9, 12–14; as of 1998, there are nineteen DAR markers, including ones installed in 1991 and 1998; six counties lack DAR markers (Maury, Lewis, and Wayne in Tennessee; Clay in Mississippi; Cobert and Lauderdale in Alabama); five of the markers are found on park land (Madison, Choctaw, and Lee in Mississippi; Lawrence and Hickman in Tennessee)

4. An Act to Provide for the Appropriation of $50,000 (48 Stat. 791), May 21, 1934.
5. Letter, S. M. Byrd to Mrs. Ferriday Byrnes, July 7, 1934 (NTPA Collection, NATR); Texas Society–DAR Homepage, (http://www.tsdar.org/kinghwy/htm; accessed October 14, 1998). The Texas DAR placed more than 120 granite boulders at five-mile intervals along the route from Mexico through San Antonio, to the Louisiana state line. The DAR paid for the historical survey, with the support of the state legislature and railroads.
6. *History of the MS-DAR,* 20, 119; Letter, S. M. Byrd to Mrs. Ferriday Byrnes, July 7, 1934. Also, according to Byrd, the group adopted a design for an iron marker (Native American head over text) to mark "crossroads and town places," exclusive of the DAR Old Trace markers.
7. *History of the MS-DAR,* 48, 74.
8. Letter, R. F. Byrnes to J. J. Byrne, May 14, 1937, NTPA Collection, NATR.
9. Letter, J. S. Mayfield to R. F. Byrnes, May 26, 1934; Verbie Lovorn Prevost, "Roane Fleming Byrnes: A Critical Biography," Ph.D. dissertation, U of Mississippi, August 1974, 159.
10. Letter, J. S. Mayfield to R. F. Byrnes, May 26, 1934.
11. Cited in draft HAER Documentation, No. MS-15, 30, NTPA Collection, NATR.
12. "Path to Parkway—the Natchez Trace," *Roadways,* October 1979, 8, 10; in Prevost, "Roane Fleming Byrnes," 160.
13. Letter, Lucille Mayfield to Officers and Members of the NTA, March 18, 1935 (NTPA Collection, NATR); letter, Lucille Mayfield to R. F. Byrnes, probably March 18, 1935 (NTPA Collection, NATR); multiple sources, cited in Prevost, "Roane Fleming Byrnes," 161.
14. Letter, NTA to the Honorable Henry D. Rainey, Speaker of the House of Representatives, D.C., July 10, 1934 (NTPA Collection, NATR).
15. "Many Delegates Attended Rally with Enthusiasm," *Natchez Democrat,* July 21, 1934 (NTPA Collection, NATR).
16. William M. Thomas, "Queen of the Trace," *Mid-South: Commercial Appeal Magazine,* September 7, 1969, 6–12.
17. "Parkway Would Aid in National Defense Program," *Louisiana Natchez Trace Extension,* September 26, 1940, n.p.; "Ancient Natchez Trace Is Viewed as Modern Atomic Highway," *New Orleans Times-Picayune,* September 4, 1953 (NTPA Collection, NATR).
18. "Bilbo Urges Mexico President to visit Mississippi 'Trace,'" *Natchez Democrat,* July 19, 1937; "High Official of Mexico will visit Natchez," *Natchez Democrat,* August 11, 1937; "Trace Celebration at Brandon Hall," *Natchez Democrat,* October 29, 1937 (NTPA Collection, NATR).
19. Memorandum, Interior Secretary to Andrew Somers, Chair-Committee on Public Lands, February 10, 1949.
20. W. F. Minor, "Annual Meeting Is Held by Natchez Trace Association," *New Orleans Times-Picayune,* October 25, 1968; W.F. Minor, "*Trace Losses* 'Its Greatest Friend,'" *New Orleans Times-Picayune,* October 20, 1970.
21. "Roane Byrnes Died Saturday, Funeral Monday," *Natchez Democrat,* October 4, 1970.
22. Prevost, 105.
23. Prevost, 112.

24. Application forms, Box 18, Byrnes Collection, U of Mississippi Library.
25. Cited in Prevost, "Roane Fleming Byrnes,"173.
26. It was produced over six months by Mabel Lane (1911–92), a photographer and colorist who ran a studio in Natchez. The Historic Natchez Foundation owns the Mabel Lane Collection (610,000 photographs/negatives), which include the Old Trace images. The photo mural, removed from Ravennaside in 1997, is on loan to the park. Interview with Mimi Miller, Historic Natchez Foundation, 1997.
27. Memorandum, notes of a conference on Natchez Trace Survey, May 17, 1934 (NTPA Collection, NATR).
28. "Mrs. Roane F. Byrnes Presented with City Bank and Trust Co Trophy . . . ," *Natchez Democrat,* January 2, 1937 (NTPA Collection, NATR).
29. U.S. Department of the Interior, National Park Service, Natchez Trace Parkway Survey, Sen. Doc. 148, 76th Cong: 3rd sess.; published by GPO in 1941, 145–46.
30. Memorandum for files, May 17, 1934, 4–6 (NTPA Collection, NATR). No author is given for this memorandum, but it was based on a conference in Washington, D.C., attended by John Nolen, H. J. Spelman, Thomas G. Vint, V. E. Chatelain, Herman Kohn, E. S. Zimmer, and G. J. Albrecht.
31. "Mrs. Byrnes Elected Natchez Trace Association President," *New Orleans Times-Picayune,* October 27, 1967; in Prevost, "Roane Fleming Byrnes," 206.
32. Letter, R. F. Byrnes to Jackson Price, June 4, 1968; in Prevost, "Roane Fleming Byrnes," 213.
33. Thomas, "Queen of the Trace," 7.
34. "Natchez Trace Group to Meet Here Oct. 15," *Tupelo Daily Journal,* July 10, 1970.

CHAPTER 8

ROADSIDE LUXURY: URBAN HOTELS AND MODERN STREETS ALONG THE DIXIE HIGHWAY

R. STEPHEN SENNOTT

When most aficionados of roadside history and culture think of early-twentieth-century travel, they contemplate motels and motor courts found along the road in small towns. It is the nostalgic mom-and-pop operation, inexpensive and with minimum comforts, that is studied and written about. However, there was also a market in mid- and large-size cities for the middle-class tourist. Accommodations in these cities were not merely a stop along the way; they were just as often a destination. The Dixie Highway Association's headquarters in the Patten Hotel is but one example. As much a part of the southern roadside as motor courts, the urban hotel provided modern amenities and comfort to tourists as they traveled from city to city. Designed by important architects during the City Beautiful Movement, these well-appointed hotels also showed the broad influence of Chicago architects and hotel developers along the Dixie Highway.

"You arrive at night and leave early in the morning and all you see is one street driving in, and another going out, and a lobby, dining-room and a bedroom or two at the hotel."

—Emily Post

By Motor to the Golden Gate, 1916

In one sentence Emily Post suggested the linear sense of mobility, compressed sense of time, and narrowed windshield view experienced by her contemporary motoring travelers as they moved between distant places, along city streets, and among hotel rooms early in the motor age. By the late 1920s, when the Good Roads Movement had linked many established urban centers and farm villages, or had brought together citizens of the North and the South, the continuous concrete road and affordable automobile had joined with telephone lines and electricity as symbols of progress in an industrialized nation. Originally designating Chicago as the northernmost point, and at one time dubbing the north-south route the "Hoosier Land-to-Dixie Highway," the Dixie Highway Association began meetings in Chattanooga, Tennessee, in April 1915 to determine its inclusive and winding route along two geographical divisions that eventually carried motorists between Sault Ste. Marie, Michigan, and Miami, Florida. By 1925 the Dixie Highway system incorporated a total of 5,786 miles, and the affordability of the automobile and the increased mobility of auto tourists significantly increased the demand for overnight accommodations.[1] In many downtown districts, developers and architects responded with the erection of modern urban hotels that catered not only to local citizens, but increasingly to motoring Dixie Highway tourists seeking a comfortable overnight place for rest.[2] Even while the railroads provided a reliable timetable along well-established passenger routes, many travelers and tourists soon preferred the adventure, flexibility, and freedom of individual automobile travel as they motored from state to state along two significant early national highways, the Lincoln Highway, whose association began in 1913, and the Dixie Highway.[3] The Dixie Highway Association was headquartered in one of Chattanooga's premier urban hotels, the Patten Hotel, between 1915 and 1927. This was an era when higher standards of comfort, different travel needs, and the widespread interstate mobility of automobile tourists reshaped the urban hotel.

Although older railroad hotels remained in business in many towns and cities where modern automobile highways were routed, this early generation of motor travelers and tourists motivated the hotel industry and its chosen architects to improve substantially the quality of their overnight stays. What might be called an urban hotel for motorists provided in its style and form, layout of rooms and modern amenities, skyline scale, and urban location a comfortable and desirable overnight stay that differed entirely from the popular "gypsying" or auto camping that inspired so many motorists seeking freedom along the open road. In his study of automobilists and the evolution of roadside lodging, Warren James Belasco assessed the many differences between the hotel and the auto camp experience during the early motor age. Belasco's book explains much about the habits and desires of early motorists who sought the adventure of the auto camp over the formality and schedules of the urban hotel.[4] Even though adventurous autocampers preferred the rustic outdoors, other motorists were not as troubled by the formality, dress codes, meal schedules, and urban setting that characterized the urban hotel experience. Despite the run-down character of some old hotels and the sometime-indiscrete habits of guests at these historically male institutions, new managers provided up-to-date amenities to improve the hotel setting for their new types of guests, the automobile tourists. In cities that could afford them, urban hotels for motorists were significantly reshaped from the late-nineteenth-century railroad-era forms and amenities to take advantage of new architectural forms and mechanical systems. For many Dixie Highway motorists the improved urban hotel made automobile touring a rewarding and relaxing way to visit new cities and enjoy regional scenic areas as they traveled between the North and the South. Built in relatively large numbers during the 1910s and 1920s, urban hotels, in their floor plans and ornament, reveal much about the impact of automobile travel and urbanism in the South. The visual record and architectural history of the urban hotels built along the Dixie Highway provide lasting images of quiet luxury, pleasurable travel, and tasteful urbanism more often associated with northern cities such as Chicago or Detroit than with Chattanooga and the South.

The impact of the automobile and highway travel on southern culture and its cities early in the motor age has been examined by Blaine Brownell, who concluded that while the automobile was considered a mixed blessing, it was nonetheless a "full-fledged symbol of modernity."[5] Noteworthy to Emily Post and others who noticed how rapidly the automobile was reshaping American life and its cities, the urban hotel was a resting place away from home. It was a sometimes formal but usually well appointed and deliberately arranged series of public and private spaces. From inside spacious public lobbies and dining rooms or from one of hundreds of upper-floor private guest rooms, visitors to the urban hotel could observe the modern, motorized city outside and, when fortunate, the surrounding natural scenery.[6]

Changing notions of comfort affect complex and contradictory expectations of roadside room and board, from the early medieval inn to the twentieth-century urban tourist hotel. The urban hotel's visibility in tourist postcard imagery, magazine advertisements, illustrated architectural publications, and travel literature in the early-twentieth-century urban-motoring landscape argue that the urban tourist hotel shaped the expectations and habits of motoring tourists during the early years of interstate travel along the Dixie Highway. Paralleled by the increasing affordability of the automobile and an expanding auto tourism industry, the businessmen and political leaders of the Dixie Highway Association and the Good Roads Movement dispersed the highway's idealistic and practical goals, first articulated in Chattanooga, Tennessee. Beyond the obvious economic and agricultural market benefits tied to improved roads, the Dixie Highway Association believed its "great artery of travel" would allow people of all classes and regions "to travel about and see how the rest of the people were living." Promoters believed that access to the historic sites, a healthy climate, modern and beautiful cities, and scenic districts in the South would promote tourism in both directions and thus enhance "the value of the exchange and personal friendships brought about by reason of the closer touch and the better understanding that naturally follows."[7] The Knoxville Automobile Club noted that their city's fortunate location amid regional highways and other towns meant that tourists often spent one or more nights in the city. More to the point, the club noted that touring automobiles carried between four and eight passengers, each of whom spent money in Knoxville's retail district to cover hotel, garage, and auto supply needs.[8]

FIG. 8.1. Postcard view of Knoxville, Tennessee. Lake County Discovery Museum, Curt Teich Postcard Archives.

The popularity of automobile tourism spawned all kinds of tourist lodgings, from free autocamps to moderately expensive urban hotels.[9] Much has been written about the history of rustic autocamps, places near towns and in open fields alongside the road where motorists could camp while touring scenic areas by automobile. From this desire for adventure and independence, rustic outdoor travel and camping, and freedom from "the hotel question, the bane of the automobile tourist," towns and entrepreneurs provided overnight places for auto tourists to set up their campsite, prepare their meals over an open fire, and enjoy the company of fellow autocampers.[10] Portrayed in John Steinbeck's *Grapes of Wrath* and director John Ford's 1940 film version, the municipal autocamp had become a potent symbol of the United States during the Depression. The previous decade had supported hundreds of auto-tourist camps set up on the outskirts of towns and cities along the Dixie and contemporary highways. Writing for the *Dixie Highway* in August 1921, Cole Morgan described Chattanooga's "modern auto tourist camp" as a "model of comfort, convenience and efficiency" in the city's Jackson Park, which provided parking and camping space sufficient for up to two hundred cars at a time. Morgan's description of the park's convenient location, with easy access to the city's retail district, nearby highways, and automobile services, narrates the civic pride associated with this site, managed and maintained so that "men, women and children of the highest refinement need not hesitate to take advantage of the facilities."[11] The urban hotel offered modern indoor comforts for travelers unsure of the safety and comfort of the camps or motor courts. Mass-produced picture postcards, family photo albums, early travel photographs, and American fiction and popular literature provide a visual and textual record of the idiosyncratic urban hotels, tourist cabins, and autocamps that were marked and advertised in highway guidebooks, automobile club guides and service station maps, and automobile blue books.[12] As the Dixie Highway's history and cultural landscape demonstrate, standards for luxury and tourist accommodations have been altered over time by the touring public's evolving demands for good roads, speed and access, comfortable lodgings, and any number of related commercial motivations and collective expectations.[13]

With its historical precedent linked to the medieval inn and country tavern, the urban hotel was made distinctive by its impressive size, large number of guest rooms, an array of lavish public rooms for dining and entertainment, and related spaces that might include theaters, ballrooms, private suites, and parlors.[14] As the United States expanded westward in the nineteenth century, hotels associated with the railroads were among the first substantial commercial buildings to be constructed. Mid-nineteenth-century developers and architects in cities such as Philadelphia, Boston, New York, Baltimore, and Washington had previously established basic design guidelines and decorative standards for urban hotels. In Boston in 1827, for example, the architect Isaiah Rogers designed the Tremont House, monumental for its time, and visited and noted by Charles Dickens in his 1842 *American Notes*.[15] Believed to be the first modern hotel in the United States, the 170-room Tremont House established significant architectural features that would continue to define the interior of urban luxury hotel into the early twentieth century: many private rooms arranged on long interior corridors, numerous stairways, well-furnished public rooms, a spacious and ornamented formal dining room, up-to-date plumbing and mechanical systems, and a monumental entrance lobby.[16] Facing the main street in a fashionable part of Boston, the Tremont's granite exterior followed the Greek Revival tastes of the time.[17] Further west, the architect Thomas Walsh, writing in 1863 in the *Builder* about U.S. urban hotels and his own seven-story Lindell Hotel in St. Louis, characterized these elaborately finished, richly ornamented hotels as "fine specimens of architectural taste."[18] Walsh's comprehensive descriptions and their boastful tone indicate the fierce competition between hotels and cities as they sought to be the largest, tallest, most beautiful, most modern, and most comfortable hostelries available. Among the important hotels of the late nineteenth century, Chicago's internationally known Auditorium Building (built in 1889) by Dankmar Adler and Louis Sullivan contained a four-hundred-room hotel and represented the epitome of urbanism and luxury.[19]

As an inherited building type, the twentieth-century urban hotel was expected to present a large and lavishly decorated façade in a fashionable architectural style. The urban hotel frequently held a prominent place alongside church spires and early skyscrapers in the city's skyline. Visitors and local citizens remarked upon the modern hotel's streetscape size, huge guest capacity and its modern furnishings. In tourist guide books and automobile club publications, urban hotels such as those designed in New York by the influential architectural firm, McKim, Mead and White, were ranked highest when they were the biggest, most elegant, and most expensive. Exemplified in New York by hotel architects Warren and Wetmore's 1910 Ritz-Carlton and 1914 Biltmore Hotel, or in Chicago by Holabird and Roche's 1909 LaSalle Hotel and the Congress Hotel (Clinton J. Warren, 1893; and Holabird and Roche, 1901 and 1907), American urban hotels were praised for their enormous expense, large dimensions, modern plumbing or other mechanical improvements, and a predictably elaborate and costly program of hugely spacious dining rooms, ballrooms, lobbies, and parlors (fig. 8.2).

Following this pattern along the Dixie Highway, the leading hotels were situated along the prominent avenues and boulevards or at prestigious central business district street corners of the South. Their facilities surely met expectations of southward-bound northern tourists accustomed to the likes of Chicago's elegant Blackstone Hotel. Smaller urban hotels located in mid-sized cities along the Dixie Highway fulfilled similar but less grandiose architectural expectations. Urban hotels like the Chattanooga's Park Hotel situated in a central business district best represent the architectural features of this architect-designed building type: prominent place in the skyline, stylish ornament and large entrance on principal facades, and a strong streetscape presence (fig. 8.3). While similarly-scaled hotels were built in Knoxville, Indianapolis, or Nashville, their advertising language and design imagery were purposefully imitating the much larger and well-known hotels in Chicago or New York.[20]

FIG. 8.2. Hotel LaSalle, Chicago, Illinois. Ryerson and Burnham Library, Art Institute of Chicago.

Situated near Tennessee's scenic Lookout Mountain, historic sites, and other sightseeing attractions, the industrial city of Chattanooga is near the midpoint of the Dixie Highway and served as the highway's headquarters. Strategically located along major transportation routes, Chattanooga was an important stopover for travelers

FIG. 8.3. Park Hotel, Chattanooga, Tennessee. Lake County Discovery Museum, Curt Teich Postcard Archives.

even before the advent of the Dixie Highway. The Patten Hotel and the Read House hotel exemplify the Dixie Highway urban hotel's quintessential architectural features. To attract overnight motoring tourists, advertisements in *Hotel Monthly,* the *Dixie Highway,* and highway guidebooks emphasized the Patten's and the Read

House's safety, convenience, and comfort, large numbers of guest rooms, abundant entertainment rooms, and extensive modern amenities (fig. 8.4). Advertising copy for these and other competing Dixie Highway hotels reveal what managers believed mattered most to tourists as they motored through their cities. In Lexington, Kentucky, for example, the Lafayette's manager and president, L. B. Shouse, claimed that his hotel was the "Best Hotel in the South" for its "Two Million Dollars in Solid Comfort," that included fireproof construction and an adjacent fireproof garage. In Chattanooga the Hotel Patten's president and manager described the hotel as "Modern" and "Fire-Proof," with "Home Comforts" and "Excellent Cuisine," while pointing out two other prominent Dixie Highway hotels under the same management in Savannah, Georgia, and Jacksonville, Florida.[21] What was intended to make these advertisements' hard-sell copy so persuasive, however, were the architectural images that depict these monumental hotels at the intersection of two prominent downtown streets. Usually shown with an orderly representation of pedestrians and automobiles, these tall and modern hotel towers—particularly the Patten Hotel—resemble and inevitably refer to the magnificent Chicago School skyscrapers in Chicago, New York, Buffalo, Pittsburgh, and Cleveland that were designed by prominent Chicago architects. Whether relatively small or large in scale, the basic form of Chicago School building consisted of three principal compositional parts: a pedestrian-scaled and ornamented base of retail storefronts, a vertically-accented central stack of offices made visible by the expressed and ornamented skeletal steel frame, and a prominent and ornamented top story and roof cornice.[22] The impact of this influential Chicago School urban form was felt in cities along the Dixie Highway, as shown by the Patten Hotel and the Park Hotel in Chattanooga. The Chicago School architectural office of Holabird and Roche made significant contributions to the national inventory of expensively furnished, modern luxury urban hotels. Its design for the Read House hotel demonstrates how the modern features and urbanism of the Chicago School tall building were adaptable for urban hotels along the Dixie Highway.

Built at great expense and "Catering to Tourists" during the 1920s motor age heyday of the urban luxury hotel, the Read House was designed by the prolific Chicago architectural firm of Holabird and Roche, known nationally at the time for their elegant skyscrapers and well-equipped commercial and civic structures of the modern city.[23] Their firm designed numerous urban hotels, among them structures that Dixie Highway hotel managers and owners would have seen published in contemporary magazines and trade publications. Two examples are the Hotel LaSalle, a "metropolitan hotel of the highest type,"[24] and the Congress Hotel and Annex.[25] Built at a cost of approximately 1.168 million dollars, the ten-story, 287-room Read House exhibited the characteristic features of the urban luxury hotel. These included the prominent and richly ornamented terra-cotta, limestone, and brick

FIG. 8.4. Advertisements from the *Dixie Highway*. Ohio State University Library.

façades; urban streetscape; lavish interior public spaces decorated with mahogany walls and terrazzo floors; grand entertainment rooms; monumental entrance lobby; and roof garden and terrace overlooking scenic Lookout Mountain and the downtown that would provide "amusement and recreation for the guests." Booster journalism described the Read House as "a hotel of which any city in America might be proud—a hotel of character—something different—a real temple of hospitality, where every modern convenience is installed.[26] Before selecting Holabird and Roche, Samuel Robertson Read, the son of the original Read family founders of an earlier hotel on this site, visited several architectural offices and modern hotels so that he could commission a new hotel of the highest quality by accomplished architects. When the hotel was completed in July 1926, the architects wrote of Chattanooga's hospitality and its urban imagery as reflected to travelers and local citizens by the Read House: "Similarly (referring to Georgian London), Chattanooga, now conscious of its increasing importance as a commercial and tourist center, has a modernized version of these manor houses and palaces, which is to serve as a haven of comfort and hospitality to its constant stream of visitors from all over the country."[27]

The published I-shaped floor plan makes clear how single and en suite guest rooms were arranged along straight corridors to provide comfort, privacy, views toward the scenic landscape, and modern mechanical and plumbing amenities. An elegant ballroom and six dining rooms in Chattanooga's Read House would compare favorably to those designed by these architects for Chicago's most elegant hotels. Laid out around the elegant lobby, the Read House's first floor plan shows retail shops, the main dining room, a lunch room, a drugstore, and an eight-chair barbershop, all of which suggest that these hotels were designed by and largely for a middle- and upper-class social system (fig. 8.5).[28] As motoring expanded through the South during the 1920s, other prominent Dixie Highway cities such as Atlanta could boast of urban beauty, improved linking roads, and professionally planned avenues in advertisements that often featured the newest and most elegant urban hotel as visual evidence of these claims to be modern.

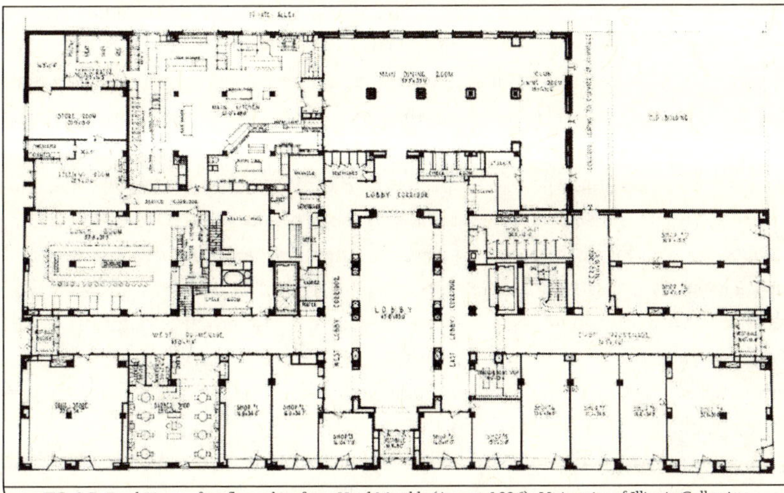

FIG. 8.5. Read House, first-floor plan, from *Hotel Monthly* (August 1926). University of Illinois Collection.

Prominent streetscapes, urban hotels, or civic buildings sometimes shaped the travel narratives of Henry James, Sinclair Lewis, Emily Post, and Theodore Dreiser, among many other observers less well known.[29] Regarding a hotel's expected comforts, Theodore Dreiser mocked his contemporaries' notions of luxury as in a Terre Haute hotel's "grill or restaurant, made to look exactly like a western architect's dream of a Burundian baronial hall."[30] As both a symbol and memory of a city once visited, the hotel marked the beginning and end of a day of motoring for tourists along the Dixie Highway. Describing road conditions along their Dixie Highway itinerary in December 1919, Natalie Newell and Gladys Ulrey closed each day's entry with the name of their chosen hotel, including the Hotel Seelbach in Louisville, the Hotel Ansley in Atlanta, and the Hotel McAllister in Miami.[31]

At the end of a day's driving, sometimes over rough roads and through inclement weather, motoring travelers may have rejected the urban hotel for its congested downtown location and distant or inadequate parking facilities.[32] But urban hotels designed during the motor age began to provide a desirable modern amenity to attract the overnight business of the motoring traveler, space in a hotel parking garage. "For the benefit and convenience of Auto Tourists," reads the 1917 advertisement in the *Dixie Highway,* "the GRAND HOTEL furnishes FREE for each car a separate fire-proof garage." Each garage had its own lock and key, permitting the owner to leave baggage and personal items in their car "absolutely free from molestation."[33] Similarly, in downtown Chicago the Hotel LaSalle offered parking for its guests in a five-story concrete garage, all designed by Holabird and Roche.[34] When considering the highly competitive quality of the hotel business during the early years of U.S. middle class tourism, one finds that hotels with new automobile garages adjacent to their buildings had an advantage over competitors. The architectural and promotional evidence confirms that a significant portion of automobile tourists from northern cities sought out familiar levels of comfort and luxury when they motored through unfamiliar southern cities along the Dixie Highway. It is evident, too, that urban hotels in cities along the Dixie Highway directed advertisements and postcard views to traveling motorists whom preferred hotels with modern equipment and comfortable public spaces (fig. 8.6). With Chicago and its City Beautiful Movement's sense of urbanism linked by spur routes to the Dixie and Lincoln highways, motoring tourists were invited to enjoy the comforts of its largest and most elegant hotels. Elegant in exterior style, perceived as modern in design, and equipped with up-to-date interior amenities, urban residential and tourist hotels such as Chicago's Shoreland Hotel and Palmer House generously catered to the needs of visiting motorists while simultaneously delivering powerful images of luxurious public space and refined urbanism.

Concurrent with the development of early interstate highways like the Dixie Highway, the City Beautiful Movement was fulfilling its planning promises in cities

FIG. 8.6. Views of and from the Park Hotel, Chattanooga, Tennessee. Lake County Discovery Museum, Curt Teich Postcard Archives.

like Chicago, Miami, and Minneapolis, as well as in smaller cities located along these early highways. In the wake of their influential 1909 *Plan of Chicago,* architects Daniel Burnham and Edward H. Bennett envisioned organized and motorized dream cities.[35] Expansive boulevards and homogeneous architecture, they believed, would replace congested city streets and ugly skylines. In addition to its famous urban plan and boulevard-lined parks, Chicago and its famous skyscraper architects also defined the highest standards of urban order and architectural beauty. Given their large scale, enormous expense, modern equipment, lavish public spaces, dignified streetscape presence, and elegantly decorated facades, Chicago School skyscrapers resembled and gave form to the urban hotel as a building type that imposed itself upon the urban landscape.

Insofar as the skyscraper defined urbanism and modernity in twentieth-century U.S. cities of any size, the urban hotel and its associated automobile culture conveyed a similar sense of modernity. Built in relatively large numbers during the 1910s and 1920s, urban hotels in their floor plans, elevations and ornament reveal something about the impact of automobile travel and urbanism in the South. Large and fashionable hotels such as those in South Bend and Indianapolis, Indiana, were situated along the prominent boulevards or at the prestigious corners of central business districts, and their facilities surely met expectations of Florida-bound northern tourists accustomed to the likes of Chicago's or Detroit's large and elegant urban hotels. From the North to the South, the traditional urban sense of place and scale was increasingly transformed by the cultural impact of the popular, mass-produced automobile; the different set of expectations carried southward with northern tourists; and the architectural impact of the urban hotel and skyscraper.

Midway between the northern origins of the Dixie Highway and the southern Florida destinations, the city of Chattanooga supported an active automobile culture and attracted motorists to its scenic tourist attractions at Lookout Mountain. Tourists might choose to stay in the five-story Grand Hotel, which was advertised to be a reasonable tourist hotel for motorists who preferred something downtown with comfortable accommodations (fig. 8.7). Medium-scale hotels such as the Park Hotel existed in many smaller cities when the first waves of automobile tourism and highway designations channeled visitors through downtown districts. Guidebooks listed hotels such as these as comfortable and sufficient for overnight or short-stay automobile tourists. While this kind of urban hotel did not by design accommodate the automobile tourist, its proprietors benefited from its downtown location on or near the newly posted highway routes. As postcard views indicated, the Hotel Patten exhibited most of the definitive architectural features of the Chicago School skyscraper that sometimes characterized modern and well-equipped tourist hotels, including the ornamental features of the skyscraper-like facade, the comfortable public spaces, the entertainment rooms, and the enormous size of this structure in its urban setting.

Arriving in Jacksonville or the resort landscape of Miami, upper-midwestern tourists had reached their destination.[36] Having utilized the urban hotels along the Dixie Highway for overnight stops along the way, the northerners were now pre-

FIG. 8.7. Grand Hotel, Chattanooga, Tennessee. Lake County Discovery Museum, Curt Teich Postcard Archives.

pared to stay for longer periods to pursue their leisure-time activities under the Florida sun. The extravagance and fantasy that frequently characterized Florida resort hotel settings fulfilled the tourist expectations as promoted by Carl Fisher and the expansive Florida tourist industry. These luxury hotels emphasized both the vacation landscape of Florida and the urbanism and modernity that were associated with major northern cities. It should also be noted that the urban hotels were not built in a vacuum. Miami Beach, at the end of the Dixie Highway, illustrates this. The urbanism of the early motor age, best characterized as those spaces and structures inspired and designed by the City Beautiful architects and planners, had been brought to the streets and urban spaces of Carl Fisher's Dixie Highway destination, Miami Beach. The work of Chicago planner and architect Edward H. Bennett in Miami Beach reiterates that the orderly urbanism of the North had taken hold in the relatively untamed tourist landscapes of Florida.

In May 1913, four years after the publication of Daniel Burnham's and Edward H. Bennett's *Plan of Chicago,* Bennett joined visiting architects, city planners, and municipal leaders in Chicago for the Fifth National Conference on City Planning. Literally surrounded by the rapidly growing numbers of motorized vehicles, the urban planners were seeking to define new concepts of urban mobility and to manage pedestrian and vehicular congestion. Typically City Beautiful architects sought expressions of urban beauty combined with orderly pedestrian and vehicle movement, utilizing the street and its fixtures to ornament cities with fountains, paved and decorated plazas, and beautiful light standards. Sometimes incorporated by national highway routes such as the Dixie and Lincoln highways, City Beautiful street systems included encircling and radiating highways, spacious landscaped boulevards linking city parks, widened axial or diagonal thoroughfares for relieving congested urban centers, and tree-lined residential streets. Given its national reputation for street planning and comprehensive city plans, Bennett's planning and architectural firm, Bennett, Parsons, and Frost, was asked to participate in Miami as professional advisers (not the jurors) for two design competitions conducted by the Biscayne Boulevard Association in 1926, one year before the Dixie Highway Association disbanded, and when improved roads were more the norm, especially in downtown commercial districts.[37]

One design competition was for a street traffic-signal tower and standards, street lighting, and street-name signs. The second design competition was for a filling station, an increasingly important building type because of its indispensability to travelers, its prominent corner siting,[38] and its explosive growth in numbers, paralleling the increased amount of motorized vehicles. According to the competition's program, architects were invited to submit designs appropriate to the recently widened and improved 3.5-mile boulevard running north from Miami, through a well-to-do residential area, and eventually connecting with the north-south Dixie Highway. Competitors were expected to submit designs that added a sense of urbanism and dignity to the boulevard, namely a wide street planted with royal palms that would serve as a gateway for Dixie Highway motorists traveling to Miami beginning in January 1927. The Biscayne Boulevard competition results were published in the *National Municipal Review, Architectural Record,* and *American City,* indicating the importance of street fixtures and carefully designed roadside landscapes to the City Beautiful planners and resort-area developers during the early motor age.[39] It is significant that the Dixie Highway stood at one end of the projected boulevard, for this City Beautiful space represents the ultimate destination for northern urbanism, namely the Florida tourist landscape.

According to published reports, the Biscayne Boulevard Association obtained highly satisfactory designs for filling stations and light standards. The competition produced ninety-three designs that were submitted from all over the United States, as well as from France, Germany, and England. The jurors and boulevard association believed that the winning architects had created appropriate standards of beauty for light-fixture designs, as shown by the published prizewinning designs. In their view this relatively small-scale street ornament represented in essence a highly desired sense of urbanism for the 1920s U.S. city. The project's boulevards and roads enhanced the tourist landscape just as the downtown hotel enhanced the urban and motorized landscape of Dixie Highway cities from Chicago to Miami.

When the Dixie Highway Association ended its work in 1927, the urban luxury hotel had proliferated on major avenues in small and large cities located along the highway. Familiar with modern hotel management, Robert Ludy characterized twentieth-century hotels as "a palatial but comfortable home for the traveler accustomed to expect luxury."[40] The modern urban tourist hotel satisfied these expectations with its monumental scale, fashionable architectural ornament, lavishly decorated restaurants and parlors, energetic staff, modern amenities, and comfortable guest rooms. Prominent on the skyline and visible to motoring tourists and local citizens as they approached their downtowns, the urban hotel represented in scale and style a desired sense of civic enlightenment, urban beauty, and modern convenience. At one end of the Dixie Highway, the resort landscapes of Florida frequently provided middle-class and wealthy tourists with extravagant images of leisure and natural splendor. Adopting the fashionable Art Deco and Mediterranean styles to the sun-drenched climate, the developers and architects of the luxury hotels and resorts designed fabulous places along the ocean beaches, accommodations that shared little with the elegant but smaller urban hotels of Indiana, Tennessee, or Georgia. Motoring through those states with relative haste, the southbound Dixie Highway tourists from the North had arrived at their vacation destination, characterized in 1926 as "a region of winter resorts, some gay and some restful; a place to which people come to escape from everything that reminds them of the North."[41] With a change of seasons, northbound Dixie Highway tourists from the South would motor to the cooler summer vacation landscapes of the Atlantic and upper Midwest states. After leaving behind the Florida resort hotels and after driving past the rural automobile campsites, they would have the opportunity to rest overnight in the comfortable and well-appointed urban tourist hotel, an accommodating building type that represented urbanism, modernity, and luxury along the avenues and boulevards in the twentieth-century city reshaped and recreated by the automobile and the needs of motoring Dixie Highway travelers.[42]

NOTES

Thanks are owed Christopher Durso, Richter Apprentice Scholar (Summer 1998) at Lake Forest College, who assisted with research for the original conference paper, this chapter, and for a related freshman studies course, "The American Roadside." The students were vigorously engaged in the materials. Amanda Mertens, peer teacher for the course, provided insightful comments for revision. Thanks are also due Debra Gust of the Curt Teich Postcard Archives, Lake County Museum (Wauconda, Illinois), for her assistance in locating Dixie Highway postcards; Philip Langdon, for his thoughtful contributions to revisions; Amy Ballmer and Mary Woolever of the Ryerson and Burnham Libraries in the Art Institute of Chicago; Graham Walden, Mel Ankeny, and Jennifer Kuehn of the Ohio State University Library System; Matthew Cook, head of library (Graham Resource Center) in the College of Architecture at Illinois Institute of Technology; and Claudette Stager of the Tennessee Historical Commission.

1. Phil T. Colgrove, "The Story of the Dixie Highway," *Dixie Highway* 11 (March 1925): 1. Colgrove was vice president of the Dixie Highway Association and president of the Michigan State Good Roads Association.
2. Additional background information supporting this study has been gleaned from Martha Carver, "Driving the Dixie: The Development of the Dixie Highway Corridor," *SCA Journal* 13 (Fall–Winter 1994–95): 11–18, and Pam S. Ecker, "The Historic Context of the Dixie Highway," available at http://www.us-highways.com/tzimm/ecker.htm (accessed May 21, 1998).
3. Business leaders, politicians, automobile enthusiasts, road builders, and transportation planners joined the Lincoln Highway Association and the Dixie Highway Association to collaborate with local leaders to designate the route for the transcontinental Lincoln Highway and the north-south Dixie Highway. See Drake Hokanson, *The Lincoln Highway: Main Street across America* (Iowa City: Iowa State UP, 1988). See also *The Lincoln Highway* (N.p.: Lincoln Highway Association, 1935), an early highway guidebook.
4. Warren James Belasco, *Americans on the Road: From Autocamp to Motel, 1910–1945* (Cambridge: MIT P, 1979). This well-documented book provides the historical and social context for the early motor age. See "Heading South without Getting Sidetracked" in this volume, where the hotel and autocamp experiences are compared. Belasco usefully

categorizes the hotel type to include the summer resort hotel, the commercial urban hotel, and the small town drummers' hotel, all designed to accommodate railroad tourists.

5. Blaine A. Brownell, "A Symbol of Modernity: Attitudes toward the Automobile in Southern Cities in the 1920s," *American Quarterly* 24 (March 1972): 27. For further analysis of automobility, good roads, and tourism in the South, see Howard Lawrence Preston, *Dirt Roads to Dixie: Accessibility and Modernization in the South, 1885–1935* (Knoxville: U of Tennessee P, 1991).
6. See David R. Goldfield, *Cotton Fields and Skyscrapers: Southern City and Region* (Baltimore: Johns Hopkins UP, 1982), for an assessment of how the southern cities often lagged behind the progress made by their northern neighbors. Among other modern technologies, the automobile, the electric light bulb, and the telephone had broad social implications. For relevant discussion of contemporary technologies, see David E. Nye, *Electrifying America: Social Meanings of a New Technology, 1880–1940* (Cambridge: MIT P, 1990), and Claude S. Fischer, *America Calling: A Social History of the Telephone to 1940* (Berkeley: U of California P, 1992).
7. Colgrove, "Story of the Dixie Highway," 2. During World War I the Dixie Highway had importance for military planning and national cohesion; at that time the Civil War still haunted southern culture. See also "Past, Present, and Future of the Dixie Highway," *Dixie Highway* 3 (1917): 1–7.
8. "Nearly Half Million Spent by Tourists Annually When They Travel through Knoxville," *Dixie Highway* 5 (November 1921): 20.
9. On early auto camping and touring, see Belasco, *Americans on the Road.*
10. From Zoe A. Tilghman, "An Automobile Camping Trip," *Woman's Home Companion,* June 1917, 31, quoted in Belasco, *Americans on the Road,* 41.
11. Cole E. Morgan, "Chattanooga Establishing Modern Auto Tourist Camp, *Dixie Highway* 5 (August 1921): 5–7. Morgan was secretary to the Chattanooga Automobile Club, which joined with City Park Department to provide a shelter house, a community house, men's and women's retiring rooms, large kitchens, lavatories, electric lights, shower baths, dining pavilion, and heat in a park setting that included mature trees, benches, and playground equipment.
12. In 1996 the Hermon Dunlap Smith Center for the History of Cartography at the Newberry Library in Chicago organized an exhibition, "Paper Trails: Maps, Highways and American Journeys in the Twentieth Century," curated by James Akerman.
13. For historical background related to twentieth-century tourism, see John A. Jakle, *The Tourist: Travel in Twentieth Century North America* (Lincoln: U of Nebraska P, 1985), and Dean MacCannell, *The Tourist: A New Theory of the Leisure Class* (New York: Schoken Books, 1976).
14. Any study of building types begins with Nikolaus Pevsner, *A History of Building Types,* A. W. Mellon Lectures in the Fine Arts (Princeton: Princeton UP, 1976). See pp. 169–192 in Pevsner, "Hotels," for a lengthy history and numerous illustrations of nineteenth- and twentieth-century U.S. hotels.
15. Isaiah Rogers, his Tremont House, and Boston's Greek Revival architecture are examined in Talbot Hamlin, *Greek Revival Architecture in America* (London: Oxford UP, 1944). By the

early nineteenth century, U.S. architects including Asher Benjamin of Boston and Benjamin Latrobe of Richmond, Virginia, were associated with impressive urban hotels designed with domes and columns of the fashionable Greek Revival style. Hamlin calls the Tremont House the first modern hotel in the United States. For excerpts from Dickens, see p. 186 of Hamlin.

16. Hamlin, *Greek Revival Architecture,* 111–14.
17. Pevsner, *History of Building Type,* 175.
18. Ibid., 180. The Lindell could accommodate twelve hundred guests.
19. For more on Chicago's hotels, see C. W. Westfall, "From Homes to Towers: A Century of Chicago's Best Hotels and Tall Apartment Buildings," *Chicago Architecture 1872–1922: Birth of a Metropolis,* ed. John Zukowsky (Chicago: Art Institute of Chicago, 1987), 77–89. For Sullivan, see Wim de Wit, ed., *Louis Sullivan: The Function of Ornament* (New York: W. W. Norton, 1986).
20. The Curt Teich Postcard Archives document many large and small Dixie Highway urban hotels. Looking at just Tennessee, one notices that the Grand Hotel, Hotel Northern, Hotel Ross, North Hotel, and Lookout Mountain Hotel joined the well-known Read House hotel and the Patten Hotel in Chattanooga; Knoxville had the Hotel Athens, the Colonial Hotel, the Hotel Arnold, and others; and Shelbyville had the Dixie Hotel.
21. Promotional language shown in advertisements in the December 1921 issue of the *Dixie Highway,* where J. B. Pound, the president, and John Lovell, the manager, are indicated.
22. See Carl Condit, *The Chicago School of Architecture* (Chicago: U of Chicago P, 1964); Carol Willis, *Form Follows Finance: Skyscrapers and Skylines in New York and Chicago* (New York: Princeton Architectural P, 1995); and Sarah Bradford Landau and Carl W. Condit, *Rise of the New York Skyscraper, 1865–1913* (New Haven: Yale UP, 1996).
23. This firm's history and building inventory are assessed in Robert Bruegmann, *Holabird and Roche, Holabird and Root: An Illustrated Catalog of Works, 1880–1940,* 3 vols. (New York: Garland, 1991; for the Read House, see vol. 2, cat. no. 957, 269–73.
24. See the firm's contemporary brochure, *Hotels Designed by Holabird and Roche* (unknown date), 9, which illustrates the Hotel LaSalle, Hotel Sherman, Congress Hotel and Annex, and Fort Dearborn Hotel in Chicago, among others.
25. See Westfall, 289n29, describing hotels pictured in "Hotel Association of Chicago," *Hotel Monthly* 20 (June 1912): 40–43.
26. Bruegmann, *Holabird and Roche, Holabird and Root,* 269, and citing the *Chattanooga Times,* May 29, 1925, 5. The most-detailed description was published in *Hotel Monthly* 35 (August 1926): 31–48.
27. From *Hotel Monthly* 35 (August 1926), quoted in Breugmann, 271.
28. Several scholars have written about the gendered spaces and social rituals that defined the early-twentieth-century urban hotel. See Martin Wachs, "Men, Women, and Urban Travel: The Persistence of Separate Spheres," in *The Car and the City: The Automobile, the Built Environment, and Daily Urban Life,* ed. Martin Wachs and Margaret Crawford (Ann Arbor: U of Michigan P, 1992), 86–100; Virginia Scharff, *Taking the Wheel: Women and*

the Coming of the Motor Age (Albuquerque: U of New Mexico P, 1991); and Belasco, *Americans on the Road,* 57–58.

29. Among the more-noticed motor-age travel narratives, see Henry James, *The American Scene* (Harper & Brothers, 1907); Effie Gladding, *Across the Continent by the Lincoln Highway* (New York: Brentano's, 1915); Emily Post, *By Motor to the Golden Gate* (New York: D. Appleton, 1916); Theodore Dreiser, *A Hoosier Holiday* (New York: John Lane, 1916) and *The Color of a Great City* (New York: Boni and Liveright, 1923); Sinclair Lewis, *Free Air* (New York: Grosset & Dunlap 1919); Frederic F. Van de Water, *The Family Flivvers to Frisco* (New York: D. Appleton, 1927). For discussion of these authors and U.S. tourism, see also John A. Jakle, *The Tourist: Travel in Twentieth Century North America* (Lincoln: U of Nebraska P, 1985).
30. From Theodore Dreiser, *A Hoosier Holiday,* quoted in Jakle, *The Tourist,* 255.
31. [Misses Natalie Newell and Gladys Ulrey], "Driving from Chicago to Miami in December," *Dixie Highway,* December 1919, 21–23. Traveling during daylight at a leisurely pace, they averaged 120 miles per day and traveled between seven and ten hours each day. Hotel descriptions were absent.
32. Citing local newspapers, Brownell shows that critics in large cities such as Atlanta, Birmingham, and Nashville complained about how the tall and massive urban hotels, often containing hundreds of guest rooms, contributed to downtown street, auto, and pedestrian congestion just as significantly as the similarly scaled commercial skyscrapers. See Brownell, "A Symbol of Modernity." Similarly John A. Jakle examines economic factors and high costs associated with downtown hotels. For more on the hotel, see John A. Jakle, Keith Sculle, and Jefferson S. Rogers, *The Motel in America* (Baltimore: Johns Hopkins UP, 1996), chap. 2, especially 23–31. The urban hotel existed alongside the diffusion of many building types, what Jakle usefully calls a "motel morphology" that includes hotels (commercial, resort, vacation), auto camps, tourist homes, cabin camps, cottage courts, motor courts, motor inns, and highway motels that catered to a diverse set of travelers seeking different kinds of overnight accommodations. Fire-proofing and other modern features would appeal to the motorist as well as any hotel guest, even if explicit references to automobilists' needs were absent from advertisements.
33. Advertisement, illustrated with photograph of the hotel, in *Dixie Highway,* September 1917, 21.
34. Residential hotels, apartment buildings, and hotel flats were sometimes able to offer parking in urban parking garages, a new building type, located within a few blocks or adjacent to these buildings. Holabird and Roche designed several garages, and cities such as Chicago contained many 2- or 3-story garages in commercial and residential neighborhoods. See Breugmann, *Holabird and Roche, Holabird and Root,* vol. 2, p. 362. See also R. Stephen Sennott, "Forever Inadequate to the Rising Stream: Dream Cities, Automobiles, and Urban Street Mobility in Central Chicago," in John Zukowsky, *Chicago Architecture and Design, 1923–1993: Reconfiguration of an American Metropolis* (Munich: Prestel-Verlag and the Art Institute of Chicago, 1993), 53–74.

35. See William H. Wilson, *The City Beautiful Movement* (Baltimore: Johns Hopkins UP, 1989); Daniel H. Burnham and Edward H. Bennett, *Plan of Chicago,* ed. Charles Moore (1909; reprint, New York: Princeton Architectural P, 1993).
36. Donald W. Curl, *Mizner's Florida: American Resort Architecture* (New York: Architectural History Foundation and MIT P, 1984).
37. For Bennett's career, see Joan Draper, *Edward H. Bennett: Architect and City Planner, 1874–1954* (Chicago: Art Institute of Chicago, 1982).
38. Early gas stations might be situated anywhere, but as automobile travel increased, location became paramount.
39. H. T. Frost, "Prize-Winning Designs in the Biscayne Boulevard Competitions," *American City* 35 (December 1926): 835–39; *Architectural Record* 60 (December 1926): 591; *National Municipal Review* 16 (January 1927): 5–6.
40. Robert B. Ludy, *Historic Hotels of the World, Past and Present* (Philadelphia: David McKay, 1927), 304.
41. Matlack Price, "'Mediterranean' Architecture in Florida," *Architectural Forum* 44 (January 1926): 33.
42. The world-class and famous luxury resort hotels of Florida, California, and other vacation places in the United States and Europe should be considered as a significantly different hotel type than the urban tourist hotel examined here. In addition to Curl, *Mizner's Florida,* see Paul Roth, *Living Downtown: The History of Residential Hotels in the United States* (Berkeley: U of California P, 1999); Elaine Denby, *Grand Hotels: Reality and Illusion* (London: Reaktion Books, 1998), Jeffrey Limerick et al., *America's Grand Resort Hotels* (New York: Pantheon Books, 1979); and Robert B. Ludy, *Historic Hotels of the World, Past and Present* (Philadelphia: David McKay, 1927).

CHAPTER 9

TOURIST LODGING IN THE GREAT SMOKY MOUNTAINS AND THE TRANSITION FROM REGIONAL TO NATIONAL STYLE

BLYTHE SEMMER

The Great Smoky Mountains National Park is the most visited national park in the country, so it is no surprise that hotels, motels, and motor courts sprung up near the park. Gatlinburg and Townsend, Tennessee, and major routes into the park, such as U.S. 411 and U.S. 321, experienced a building boom beginning in the 1930s. Unlike urban hotels, these accommodations had a regional style that opened up the buildings to nature. Porches and native materials were used, whether for a mid-sized hotel or an individual cabin. Like the urban hotels, the accommodation itself became a destination. Almost from the beginning, particularly in Gatlinburg, the economic reality of development pressures meant that individually owned motels and motor courts would eventually be overtaken by major chain motels and hotels with no historic or architectural ties to the region, and few of these early facilities remain. In wayfarers' quests to visit wilderness areas, they have unwittingly helped to subdue nature and promote development.

Sevier and Blount counties, Tennessee, are heirs to great natural beauty, part of which is protected in the Great Smoky Mountains National Park. The two East Tennessee counties have also inherited a more recent architectural legacy developed in response to the millions of visitors who come to enjoy the outdoors while on vacation.

Roadside architecture reflects the changing character of tourism in the Smokies, and tourist accommodations are perhaps the best expression of how locals promoted "mountain culture" to visitors through architecture. Motor courts and motels also reveal how tourists' attitudes changed over time. The first generation of automobile tourists looked for quiet, rustic flavor in their rented cabins, while post–World War II visitors opted for conveniences and modern comfort. The interaction between local proprietors' innovations and visitor desires transformed the landscape of the Smokies' resort communities several times and continues to do so today.

Sevier and Blount counties are located along the southern edge of the Great Valley of Tennessee. Both contain a portion of the national park, which spans the Unaka Range into North Carolina to the south (fig. 9.1). Visitors accessing the park from the north and west most likely do so through the gateway communities of Pigeon Forge and Gatlinburg in Sevier County or Townsend in Blount County. Tracing the development of roadside architecture in these towns and along major automobile routes into the park shows how transportation decisions dating to the park's establishment in the 1930s continue to shape the area's commercial development. The

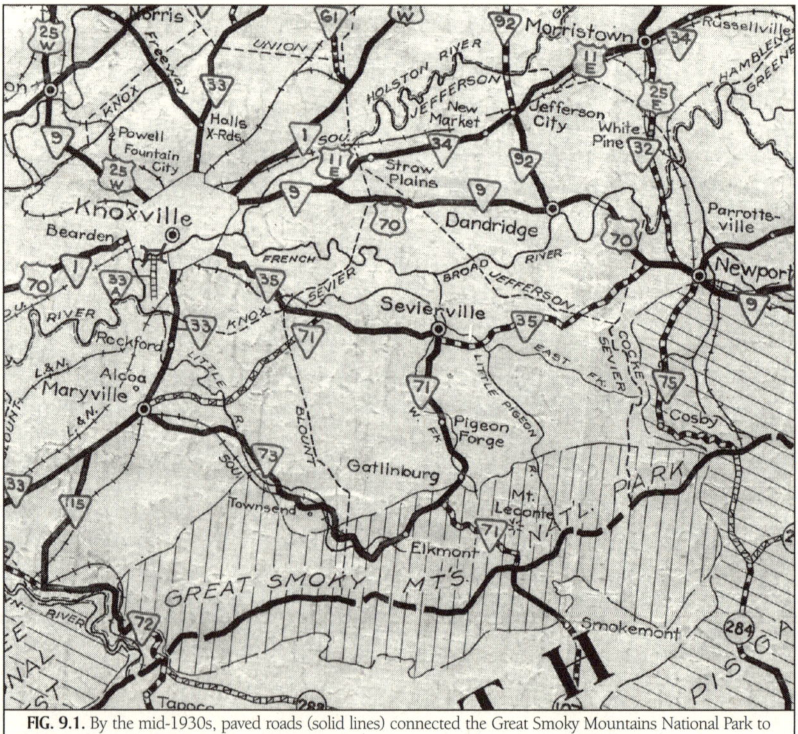

FIG. 9.1. By the mid-1930s, paved roads (solid lines) connected the Great Smoky Mountains National Park to Knoxville, Maryville, and Sevierville. Dashed lines indicate stone or gravel roads. From the Tennessee Department of Highways and Public Works, *Road Condition Map of Tennessee,* Tennessee State Library and Archives.

rapid rate of change surrounding the most visited national park and a thriving tourist destination now threatens the legacy of the motor courts and motels that lodged the first and second generations of automobile tourists.

The first roadside accommodations in the Smoky Mountains inherited an existing vernacular vocabulary of mountain tourism as much as they created new forms for the automobile age. Elites had sought the healthful mountain climate and the curative effects of spring waters, as well as social and recreational opportunities, at mountain resorts throughout the South long before the creation of national or state parks. Henry W. Lawrence, writing in the magazine *Landscape,* proposes that rural antebellum southern architecture inspired the layout of these resorts, which resembled plantations in the arrangement of a main building flanked by supporting structures. Lawrence finds that, following the Civil War, the southern resort style, inspired by Palladian-Jeffersonian tradition, became popular across the country. By the 1870s and 1880s, "a new national resort style developed: long verandas, rambling site plans, loosely connected wings, and increased concern for openness and freedom of movement, and a greater integration of buildings and grounds. Also new was the increased emphasis on courtyards, connecting walkways, and the grouping of buildings to suit the topography."[1]

Resort hotels bore little resemblance to their downtown cousins, which focused attention on the lobby and interior public spaces of the first floor. Although the processional aspects of the resort hotel were absent in tourist courts, the focus on spaces for outdoor socialization persisted in the landscaped lawns and, later, swimming pools that frequently formed the centerpieces of court and motel arrangements. Cabins or cottages surrounded central courtyards in similar fashion nationwide.

The beginning of the twentieth century brought resorts that were more accessible to their urban patrons via rail. These represented a shift from the older springs resorts in their emphasis on outdoor recreation, scenery, and exposure to mountain culture.[2] Tourism and recreation were also made available to the middle class through improved, affordable transportation, first on railroads and later in automobiles. Elkmont, a resort started by the Little River Lumber Company and served by its railroad in the Sevier County section of the Smokies, survived the creation of the park because of the efforts of summer cottage residents, who were primarily members of the Knoxville elite. The Wonderland Hotel, built there between 1912 and 1928, served as a social center for cottage residents as well as hotel guests. The two-story white weatherboard hotel was similar to other vernacular hotels of the period in that it lacked decorative details but featured a hipped roof and a wide wraparound porch on three sides.[3]

Tourists in the 1920s sought out more rustic experiences than their Victorian era predecessors. As the popularity of automobile tourism reached into the mountains, more adventurous tourists discovered that Ford Model Ts could be driven on logging

FIG. 9.2. Mountain View Hotel, Gatlinburg, Tennessee. From the author's collection.

trails. Soon motorists were visiting the balds, naturally occurring meadows on mountain peaks popular for their views. Hotels in the mountains responded to the new mode of transportation by installing gas pumps and advertising driving tours for their guests.[4]

The resort hotels of the Smokies embodied rustic functionality in their design and materials. Porches provided space for socialization, often on both primary and secondary facades.[5] Rustic elements included stone fireplaces and chimneys often made of naturally polished river stones. Both river stones and field stones were commonly used for fences, steps, posts, and general landscaping. Mountain resort structures were usually of wood-frame construction, sheathed in weatherboard or board-and-batten siding. Some mountain resort buildings were built of logs, such as the "large and commodious" eight-room lodge John Oliver had constructed in Cades Cove by 1928.[6]

The Mountain View Hotel, the first accommodation for tourists in Gatlinburg, began as a two-story, six-bedroom frame lodge built by Andrew Huff in 1916. This structure was torn down in 1922, when Huff constructed the main building of the Mountain View, which stood until 1993 (fig. 9.2). It was a three-story shingled hotel with a stone terrace overlooking the route of the parkway (State Route 71) at the present-day intersection of U.S. 321 and State Route 71/U.S. 441.[7] The Mountain View combined the structural conventions of the mountain resort hotel with a conspicuous use of local materials, creating an atmosphere that was at once refined yet evocative of the rough natural beauty of its mountain surroundings. The architecture of the major resort hotels built in Gatlinburg by the start of World War II is remarkably similar. Like the Mountain View, the Riverside Hotel (circa 1938) and the Greystone Hotel (1941, demolished 1998) were both composed of two gable-front wings connected by a central section featuring a one-story porch. Symmetrically placed paired windows, one-over-one or three-over-one, were characteristic of facades. This pattern repeated in smaller hotels as well. Another large pre–World War II hotel was the Gatlinburg Inn (1937), which departed from the gabled wing formula. It looked vaguely like a cabin, with a first floor of natural river stone and second story that appeared to be log. A one-story stone wing extended from the left side of the building.[8] A larger two-and-a-half-story building using the same architectural features was added in 1941.

Hotels and automobile-oriented accommodations such as tourist courts existed side by side in the resort community of Gatlinburg. Elsewhere around the mountains new courts and motels proliferated along highways, particularly those that were principal tourist routes toward the park. Townsend never developed a large resort hotel for the automobile age and has continued to grow along the linear corridor of the highway, as has Pigeon Forge. Maryville in Blount County and Sevierville in Sevier County continued to have town hotels to serve the needs of visitors in the county seats, but newer types of accommodations also developed on tourist routes. These businesses represented a chance for the small, local, independent proprietor to enter the tourist trade. They also helped change the traditional image of lengthy mountain vacations spent at a summer colony or resort hotel. Automobile travelers who flooded into the foothills in the 1920s, 1930s, and 1940s were interested in moving and covering territory. As a result scenic driving tours rose in popularity, and tourists looked for convenient and affordable overnight lodging at a variety of points around the boundary of the national park, although Gatlinburg remained their favorite base for exploration (fig. 9.3).

FIG. 9.3. The rustic Le Conte Creek cottages in Gatlinburg were constructed of local stone. From the author's collection.

A nationally popular trend in tourist court design was the use of regional imagery "to assure that each illusion-by-the-roadside conformed to the travel-poster image of a particular locale."[9] Tourist courts in the Smokies drew heavily on stock images of log cabins as well as mountain vernacular architecture through the use of local stone, native wood paneling, and shingles. Rustic exteriors were equipped with creature comforts inside, but owners often attempted to continue the authentic mountain appeal on the interior through the use of furniture or coverlets hand made by locals. The comforts offered in U.S. tourist courts by the 1930s were a world apart from the bare cabins that had marked the initial transition from camp to court just a few years before.

In the Smokies, domestic imagery met regional imagery in the form of the log cabin, the dwelling type that symbolized most readily the mountaineer's lifestyle. Tourist courts offered visitors the chance to stay in an individual log cabin, creating an idyllic vacation environment that combined the picturesque, rustic associations of mountain life with modern conveniences. Snug log cabins for tourists echoed the architecture of farmhouses scattered about the coves and valleys of the Smoky Mountains and created a sense that vacationers were closer to nature here than in a resort hotel. In their individual cabins by the roadside, tourists could pretend to live like pioneers without the real hardships that often accompanied life in log houses.

Shingles also commonly covered tourist courts in the 1930s and 1940s. Early on shingles appeared on the exterior of the Arrowcraft shop, which continues to sell authentic mountain crafts on Gatlinburg's main street, as do resort hotels. Other resort hotels in the area used shingle siding as well. This cladding may have been influenced by the popularity of the shingle style at elite northeastern seaside resorts around the turn of the twentieth century. When the Gatlinburg Tourist Bureau opened in 1939, it occupied a small frame building covered in hand-split shingles.[10] Shingles, stone, and unpainted woodwork were design features that harmonized with nature more readily. The early years of tourism before World War II were marked by sensitivity to the fact that natural scenery in and around the park was the major tourist draw. Builders therefore repeated mountain themes as they created an environment related to its well-known natural surroundings. A Gatlinburg newspaper editorial from the 1930s reflects this impulse: "In our little village we are judged by our works. If our town is neat and clean and we earnestly strive to keep it so then folks will see something of the character of us all. If our buildings are substantial, pleasing to the eye, and the general appeal is that of harmony with Nature then visitors will know us for what we are, creators of beauty and builders of the substantial."[11]

A 1941 account of the Gatlinburg building boom accompanied a picture of a cottage being constructed at Cox's Gateway Cabins. Logs from a dismantled house on Bluff Mountain in Blount County were used to construct the new cabin, which rested on a stone foundation and had a stone chimney and wood shingle roof. The same article refers to the popularity of constructing "quaint new cabins" out of timbers from dismantled log buildings within a forty-mile radius of Gatlinburg. The writer described a new commercial building being built of brick as "one of the few Gatlinburg structures reflecting no attempt at the rustic."[12]

Tourist court design in the nation followed trends toward streamlining in the late 1930s, embracing the modern on the exterior as well as in the interior. Although tourist courts in the Smokies and especially Gatlinburg continued to display a more regional vernacular style on the exterior, they enthusiastically advertised their modern conveniences. Automobile club recommendations and referral chains helped courts achieve recognition with tourists visiting from all over the country. In 1941 the Park Tourist Court and the French Tavern prominently advertised "modern cottages" approved by the American Automobile Association (AAA) in a Gatlinburg guide. King's Court cited its affiliation with United Motor Courts.[13] In Blount County the Derris House Motel in Townsend and the Travelers Hotel Court on U.S. 411 west of Maryville mentioned AAA approval in their ads in the 1940s (figs. 9.4 and 9.5).

FIG. 9.4. Travelers Hotel Court typified the modern impulse along Blount County highways in the 1940s and 1950s. From the author's collection.

FIG. 9.5. The Derris House Motel in Townsend, now Laughing Horse Motor Inn, shows the influence of nationally popular styles with a "Southern Colonial" twist. From the author's collection.

Motor court and motel owners in Sevier and Blount counties responded to economic change in ways similar to people in other predominantly rural farming areas in the early years of automobile tourism. Warren James Belasco points to the downturn in the agricultural economy even before the onset of the Depression as encouragement for farmers with property on tourist highways to enter the tourist court business.[14] A 1949 Tennessee State Planning Commission document on the tourist business cheerfully titled one section "Vacationers—A New Farm Crop!" and encouraged farmers to open their homes to tourists as overnight visitors or to provide cabins or cottages for vacationers near natural recreation areas. The report counseled farmers that neatly kept houses and grounds, electric lights, inside bathrooms, and "an approved water system" were necessary to attract tourists. Furthermore, "the guest who stimulates these improvements also will pay for them, and the farm family also will enjoy the new conveniences, as well as the broadened outlook stimulated by meeting and talking with the visitors."[15]

Tourist lodging in the Smokies began developing in earnest after 1930. Though visitors came to the area in anticipation of the Great Smoky Mountains National Park in the 1920s, they lodged at resort hotels or in tourist homes, which were similar to bed-and-breakfasts. One Gatlinburg writer claims that construction of boarding houses and individual log cabins began about 1930, after Congress had authorized the creation of the national park in 1926, and that tourist courts and motor courts came later as visitors demanded more and better accommodations.[16] Jeannette Greve wrote an early descriptive account of the growing resort village in 1931.

Though she mentions the Mountain View and Riverside hotels, tourist courts had not yet arrived in Gatlinburg.[17] By 1935 a roadside hostelry appeared at Tuckaleechee Cove, two miles north of Townsend. The Tuckaleechee Tourist Camp featured seven furnished cabins, as well as a tea room, dance hall, gift shop, and general store.[18] It was located next to the Standard Service Station on State Route 73, which still stands today on the old highway alignment.[19] By 1941 twenty-three tourist courts, cabins, lodges, and inns advertised in a tourist guidebook published in Gatlinburg. Of these, sixteen used the words *court, cabins,* or *cottages* in their names.[20] Four years later, Edwin J. Foscue published a study of Gatlinburg in *Economic Geography.* He reported that the arrangement of the town was basically linear, stretching more than a mile along the highway. Two roads, one up Roaring Fork Creek and the other up Baskins (Bearskins) Creek, intersected State Route 71. "The main street is lined with filling stations, modern tourist courts, restaurants, handicraft shops, and four large modern hotels, all owned and operated by local residents. Neon signs are in profusion."[21] Thus, even before the postwar tourist boom, Gatlinburg had essentially developed its tourist strip.

Before 1945 tourists in Blount County lodged in hotels and in several tourist homes along the major thoroughfares of West Broadway, Knoxville Highway, and Washington Avenue. Residential West Broadway lent itself to tourist homes, of which there were five listed in 1941 and eight in 1948. One tourist court, Coleman Court, appeared in the city directory beginning in 1941. Coleman Court was located on State Route 33, the principal road between Knoxville and Blount County and for

FIG. 9.6. Several tourist courts offered Smokies visitors a chance to stay in a log cabin. From the author's collection.

tourists heading to the Townsend entrance to the park. It was joined in 1943 by the Meadow Lark Court (demolished in 2000) and in 1948 by the Lighthouse Court. The Meadow Lark changed from "court" to "motel" by 1959, a response to modern tastes although a new motel-type building with a unified roofline was not built to accompany the name change.[22]

The Townsend area also developed along State Route 73 in the 1930s. In 1933 a property called Overlook Cottage and the Townsend Inn could be found there.[23] Two years later, a guide located Overlook Cottage "on scenic loop," a reference to the popular driving tour. That year (1935) the Tuckaleechee Tourist Camp appeared.[24] Townsend continued to develop, albeit more slowly than Gatlinburg, up to the start of World War II. "Blount County Side of the Park is 'Ready to Go,'" announced a 1941 newspaper article.[25] As plans for an improved Blount County park approach highway circulated, locals anticipated growth. The article noted that several new cottages were under construction or recently completed at Kinzel Springs, just north of Townsend on State Route 73. An accompanying photo pictured Pat and Anna's Modern Tourist Cabins, a row of board-and-batten cottages with white trim and six-over-six windows. The office was located just behind the sign, and the row of neat cabins extended to the rear, perpendicular to the road. Though the New Walland Highway (U.S. 321) was under construction close to Maryville in 1939, Townsend had to wait until after the war for road improvements.[26]

In a tourist study conducted for the Great Smoky Mountains National Park in 1949, 54 percent of car parties said they had been accommodated in cabins while visiting the park. Another 16 percent stayed in hotels, and 16 percent chose tourist homes. Nearly half of the "person-nights" spent by these travelers were spent in Gatlinburg.[27] No other location came close to Gatlinburg's popularity as a place to stay. Only 4 percent of person-nights were spent at Kinzel Springs, and a paltry 2 percent in Townsend.[28] As visitation doubled between 1947 and 1955, the demand for lodging increased proportionately. Gatlinburg continued to dominate the lodging market for park visitors. In another study, conducted in 1956, the communities on the Tennessee side of the park took in almost twice the money that North Carolina did.[29] Visitors expressed a material change in lodging preference between the two studies. In 1956, seven years after the majority of car parties said they stayed in tourist cabins, 83 percent of visitors stayed in motels, and a slightly larger percentage indicated their preference for motels. The study interpreted this difference as an indication that accommodations were lacking and that more motels should be built to satisfy visitor needs. Cabins were still included as a category in the 1956 study, but only 9 percent of respondents stated they had stayed in a cabin.[30] This rapid change in the pattern of accommodations can be attributed to the name change of many tourist courts to "motel" in the 1950s as well as the constant renovation and new construction that characterized the booming tourist industry in the Smokies after World War II.

Although Maryville, Townsend, and other communities along the major highways of Blount County could not keep pace with Gatlinburg's popularity, they also participated in the postwar tourism boom. The late 1940s witnessed significant tourist court and motel construction along the New Niles Ferry Highway (U.S. 411) and the Walland Highway (State Route 73). The 1948 Travelers Hotel Court was the first motel to be built on New Niles Ferry Highway. It was located approximately where U.S. 129 and 411 intersect today. Twenty-two connected units were arranged in a wide U-shape around a central office building. Travelers Hotel Court was an early local expression of the trend toward a unified roofline. By 1953 it had been joined by the Grove Court, Maryville Motel, and Wayside Motel on New Niles Ferry Highway, a major route through southeastern Tennessee to Atlanta. In 1959 two more new listings appeared in Maryville's city directory, the Princess Motel and the 411 Motel, which are still extant. By this time the Traveler's Hotel Court had become the Traveler's Motel and had added a restaurant.[31] Motels along U.S. 411 are artifacts of a relatively brief period during which postwar tourism and automobile travel surged before the construction of interstate highways. Although highway signs on U.S. 411 post the mileage to Atlanta, Interstate 75 now carries the vast majority of traffic from Florida to Georgia, Chattanooga, Knoxville, and the Smokies.

Similar development occurred south of Maryville on State Route 73/U.S. 321. The Park View Court was the first tourist court to arrive. It was probably built around the start of World War II. The Hacienda and Spark's Tourist Court joined it after the war. Although no photographs of the Hacienda have been found by this author, the name indicates that the national trend of identifying tourist courts with western themes and constructing them to look like western or Spanish styles of architecture may have been repeated in Blount County. Spark's Tourist Court, located next door to Joe's Restaurant, apparently evolved into the Cherokee Tourist Court in 1953, taking its name from regional themes, and later it became the 73 Highway Tourist Court. A succession of changing names may indicate changes in ownership, renovations, or both.

Col. W. C. Taylor and his wife built the Tuckaleechee Village in 1947 along Highway 73 near Townsend (fig. 9.7). The complex, which remains largely intact today, originally included an office and six double cabins. A restaurant was located just south of the office. Another building on the property, set close to the highway, served as the site of barn dances and entertainment. Closer to the park entrance, the Docks Motel was constructed in the 1950s. This low strip motel lies perpendicular to the highway. The Docks is notable for its 1950s neon Moderne sign, which still commands motorists' attention. Nearby, W. C. Derris and his wife constructed the Derris House Motel in 1946. They promised "a new and different motel . . . considered the finest in the National Park area." It featured native pine interiors, hardwood floors, and maple and knotty pine furniture. Each room in the white weatherboard structure, which had a prominent columned entry porch, was accessed from a "lounging porch." The Derris House extolled the virtues of the mountain climate and the natural recreation available nearby, but it was also accessible to Maryville and Gatlinburg and a convenient place to begin motor trips.[32] The Derris House Motel's Colonial Revival characteristics represent a departure from the rustic architecture that was common in the Smokies in favor of a nationally popular residential and commercial style. However, the building also recalled southern regional motifs, such as the two-story central porch supported by white columns or posts. Blount County's 411 Motel, built about a decade later, also added a Colonial Revival–influenced two-story porch supported by white columns during later renovations.

Gatlinburg hotels made concessions to the demands of tourists arriving by automobile, especially in the postwar period. In the late 1950s Riverside Hotel owner Bruce Whaley contracted the Gatlinburg architect Hubert Bebb to give the hotel a facelift, resulting in a strikingly modern appearance compared to its resort hotel history. The New Riverside looked more like motor lodges that were appearing along U.S. highways. Stylized stone columns and a porch covering the entire facade created a modern interpretation of the southern colonial. A cupola and shutters completed the image. The historian Ed Trout notes, "The job was typical of the fresh crop of remodelings and new construction in the 'crab orchard stone

FIG. 9.7. Tourist cabins still line U.S. 321 at Tuckaleechee Village, near Townsend in Blount County. From the author's collection.

style' (for lack of a better name), which was basically 1950s modern. It blended stone, stucco, heavy timber, large expanses of glass and clean lines. Many commercial buildings in town followed this trend."[33] The Riverside's site plan was also altered for the convenience of cars. The old rectangular parking lot was replaced with a crescent-shaped drive up to the front door. An elongated carport to the left of the entrance sheltered guests' automobiles.[34] The majority of hotels and motels in Gatlinburg paved their parking lots in the early 1950s, when the city undertook street improvements such as striping, painting crosswalks, and installing traffic lights.[35]

Both hotels and motels reacted to changing tourist traffic as the postwar period progressed. Motels advertised playgrounds as the baby boom resulted in more families traveling together. The postwar transformation was especially pronounced in Gatlinburg, the most popular Smokies resort in Tennessee. Additions and renovations began almost as soon as peace was declared, and restaurants and attractions multiplied to serve a growing clientele. Miniature golf, museums, the Sky Lift, and other attractions entertained guests when they were not seeing the sights of the national park. Postwar tourists demanded more sophistication from lodging as well as entertainment. The years 1951–54 in Gatlinburg, contends Trout, were "a watershed." The four major hotels installed telephones in guestrooms in 1951, and television arrived in 1953. Swimming pools, private baths, wall-to-wall carpeting, and air conditioners were all conveniences hotel and motel owners rushed to add (and advertise) as the 1950s boom gained momentum.[36]

Properties in Gatlinburg that had advertised their rustic cabins or handmade furniture now proclaimed updates and conveniences, reflecting the traveling public's more formal expectations of what a motel room should be. Picturesque regional variations lost ground to modernity, and conveniences, comfort, and a diversity of entertainment replaced the once-refreshing roughness of exploring by automobile. Now the more predictable comfort of standard amenities in motels assures weary tourists that they will find a "home on the road" wherever they roam.

Tourist courts and motels in Sevier and Blount counties generally followed national patterns of evolution in roadside architecture, heavily embracing rustic regional styles in their early years and taking on a more conspicuously modern appearance in the 1950s. These changes reflect the influence of national taste on the area. Change, however, also came from within. Early travelers visiting the area knew the southern Appalachian region for the independence of its residents, and Gatlinburg in particular developed a lasting reputation for being controlled by local families rather than outside entrepreneurs.[37] Local tourist court and motel owners created a distinctive atmosphere for mountain vacationers based on the appeal of the regional culture and scenery.

Although the constant cycle of exterior facelifts and new construction has removed most of the examples of early tourist lodging from the landscape of Sevier and Blount counties, a handful remain intact. City directories, guidebooks, postcards, and photographs resurrect the tourist courts and motels of the 1930s, 1940s, and 1950s. Their architectural legacy has occasionally been passed down to the modern motels, including many chains that greet Smokies visitors today. In most cases, though, modern buildings bear only a passing resemblance to the tourist courts and motor inns from which they descended. Growth continues today and threatens the last vestiges of the Smokies' early tourist accommodations. Gatlinburg's strip is now populated with national chains building higher on their small lots, although smaller local motels still survive here and there. Townsend, still reflective of the early twentieth century, now shows the greatest potential for transformative development after the recent widening of U.S. 321 through this narrow, once sleepy valley. And in Maryville and Sevierville, the last of the early- and mid-twentieth-century tourist courts and motels have gone, with the exception of a couple that remain along U.S. 411. Rapid change is the standard among businesses that cater to tourists, and each year a new crop of buildings and remodeling jobs comes forward to take the place of their predecessors. This evolution continues to demonstrate the changing nature of tourism in the Smokies while showing how national culture continues to exert its force on this once rural area.

NOTES

This essay was reprinted from *SCA Journal* 20 (Spring 2002): 14–22. Used with permission.

1. Henry W. Lawrence, "Southern Spas: Source of the American Resort Tradition," *Landscape* 27 (1983): 5–12.
2. Margaret Lynn Brown, "Smoky Mountains Story: Human Values and Environmental Transformation in a Southern Bioregion, 1900–1950," Ph.D. dissertation, U of Kentucky, 1995, 126; and Christopher Brenden Martin, "Selling the Southern Highlands: Tourism and Community Development in the Mountain South," Ph.D. dissertation, U of Tennessee, 1997, 61–62.
3. Philip Thomason and Michael Ann Williams, "Elkmont Historic District, Great Smoky Mountains National Park" (unpublished paper, U.S. Department of the Interior, National Park Service, Washington, DC, 1994), 6. Copy on file at the Tennessee Historical Commission, Nashville.
4. Brown, "Smoky Mountains Story," 130.
5. Thomason and Williams, "Elkmont Historic District," 4.
6. Durwood Dunn, *Cades Cove: Life and Death of an Appalachian Community, 1818–1937* (Knoxville: U of Tennessee P, 1985), 243, and Brown, "Smoky Mountains Story," 127.
7. Louise Durman, "Gentlemen of Gatlinburg," *Knoxville News-Sentinel,* February 15, 1987.
8. Gene Aiken, *Mountain Ways: An Album of the Smokies* (Gatlinburg, TN: Buckhorn P, 1983), 22.

9. Chester H. Liebs, *Main Street to Miracle Mile: American Roadside Architecture* (Boston: Little, Brown, 1985), 176.

10. Ed Trout and Debra Griffin, *Gatlinburg: Cinderella City* (Pigeon Forge, TN: Griffin Graphics, 1984), 105.

11. Quoted in Trout and Griffin, *Gatlinburg,* 105.

12. "Gatlinburg is A-Bustle—New Hotel, Additions, Cottages, Airport Are Being Built in Preparation for Tourist Thousands," *Knoxville News-Sentinel,* circa 1941. Undated article located in Gatlinburg history file, McClung Historical Collection, Knox County Public Library System.

13. *Guide to the Great Smoky Mountains National Park and Gatlinburg, the Park Headquarters,* 3d ed. (Gatlinburg: Gatlinburg News, 1941), 62, 58, 98.

14. Warren James Belasco, *Americans on the Road: From Autocamp to Motel, 1910–1945* (Cambridge: MIT P, 1979), 140.

15. George I. Whitlach, *Vacations: A Major Tennessee Industry* (Nashville: Tennessee State Planning Commission, 1949), 5.

16. Russell Shaw, *The Gatlinburg Story* (Gatlinburg, TN: Russell Shaw, 1979), 25.

17. Jeannette Greve, *The Story of Gatlinburg (White Oak Flats)* (1931; reprint, Gatlinburg, TN: Brazos Printing Company, 1964), 120–25.

18. George W. McCoy, *Great Smoky Mountains National Park Guide,* 2nd ed. (Asheville: Inland P, 1935), 140–42.

19. Hyldard H. Stinnett, *Nostalgia 76* (Maryville, TN: Davis Printing, 1976), 187.

20. *Guide to the Great Smoky Mountains National Park and Gatlinburg.*

21. Edwin J. Foscue, "Gatlinburg: A Mountain Community," *Economic Geography* 21 (1945): 202.

22. Maryville-Alcoa city directories, 1941, 1943, 1948, 1953, 1959.

23. *Great Smoky Mountains National Park* (1933). Pamphlet, publicity file, Great Smoky Mountains National Park Library.

24. McCoy, *Great Smoky Mountains National Park Guide,* 142.

25. Lee Davis, "Blount County Side of the Park Is 'Ready to Go,'" *Knoxville News-Sentinel,* April 6, 1941.

26. Stinnett, *Nostalgia 76,* 179.

27. The term "person-nights," as used in the study, may be clarified by this example: If a family of four stays in a hotel or other accomodations for two nights, that is equal to eight person-nights.

28. "Great Smoky Mountains National Park Travel Study" (N.p., 1949), 27–30.

29. *The Great Smoky Mountains National Park Travel Study* (N.p.: North Carolina State Highway and Public Works Commission, Tennessee State Department of Highways and Public Works, and U.S. Bureau of Public Roads, 1956).

30. Ibid., 24.

31. Maryville city directories, 1948, 1953, 1959.

32. Derris House Motel brochure, circa 1946. Townsend file, Great Smoky Mountains National Park Library.

33. Trout and Griffin, *Gatlinburg,* 121.

34. Tennessee Postcard Collection, Tennessee State Library and Archives.
35. Trout and Griffin, *Gatlinburg,* 121.
36. Ibid.
37. Local histories and guides such as Ernie Pyle, *Gatlinburg and the Great Smokies* (Gatlinburg, TN: C.C. Callaway, 1955), and Shaw's *The Gatlinburg Story* state that a handful of local families maintained ownership of most of Gatlinburg's land. Shaw writes that "a noteworthy and extraordinary fact is the large number of great grandsons who own, or manage, Gatlinburg's motels" (26). Jerome Dobson also described the role the five prominent families of Gatlinburg (the Ogles, Whaleys, Maples, Huffs, and Reagans) played in the resort's economic history. He cautioned that their tenacious hold on land applied mainly to the business district. Many rural landowners were also descendants of early settlers. As of 1968, local independent proprietors owned 77 percent of motel units. Dobson documented that local ownership was declining in the 1970s as larger, chain-affiliated properties multiplied. See Jerome Dobson, "The Changing Control of Economic Activity on the Gatlinburg, Tennessee Area, 1930–1973," Ph.D. dissertation, U of Tennessee, 1975, 78–81, 105.

CHAPTER 10

MYRTLE BEACH: MUSIC AND MOTELS

KATHERINE FULLER

By the 1950s, aided by good roads and later by the interstate system, vacationers traveled around the country from the mountains in Maine to Spanish missions in California. However, the beach was always a priority location for those seeking a respite from the city and their jobs. Some beach towns had only the beach as a draw, but more successful communities had a "special" draw, and Myrtle Beach had its "shag" music, derived from the local African American community. The beaches and the music proved an irresistible draw, and tourism became a mainstay of the town. Of course, like earlier tourists in urban areas, families driving to vacation spots, and campers in the mountains, visitors needed overnight accommodations. During that era the Populuxe motel was the equivalent of today's resort hotels: clean, modern, convenient, dependable, and located near local attractions. Katherine Fuller looks at Myrtle Beach and the design and appeal of its motels from the 1950s and 1960s and their fate.

Vacationers have flocked to Myrtle Beach, a city centered in South Carolina's Grand Strand, since the 1920s. The history of its development as a tourist destination during the first four decades of the twentieth century can be summarized as a progression from campgrounds to seaside and guest cottages to motels and hotels (fig. 10.1).

FIG. 10.1. Summer Sands, formerly the Bon Villa, circa 1960. Photo by Joel Carter, AIA, 2003.

In 1954 Hurricane Hazel cleared large parcels of land for development. Efforts afterward to attract families to the beach had an impact, but it was beach music, a form of rhythm and blues, that particularly attracted young people from the region and created a statewide phenomenon.

BEACH MUSIC

In Myrtle Beach beach music emigrated from the African American community to the mainstream. During the 1930s and 1940s African Americans were dancing the shag, a melding of the jitterbug and the lindy, and listening to swing music. Gradually a distinctive dance sound emerged with equal parts horns and rhythm, with roots in gospel and swing. As described in the 1999 book *The Beach: A History of Paradise on Earth:*

The result was irresistible, not only for the African-American musicians and beach-goers whose synergy kept the Myrtle Beach Pavilion hopping, but also for white kids in the know. But the jukeboxes and radio stations were not carrying the then forbidden fruit of this music. For one thing, its lyrical content—unabashedly sexual and bawdy—kept the songs off the radio stations altogether. The only way people living inland could catch the "underground sounds" of the era was by making the trip to the dance pavilions of Myrtle Beach. Hence the name "beach music."[1]

FIG. 10.2. The Beverly circa 1971, with the Pavilion Amusement Park in the background. Photo by the author, 2001.

Lodging during that time was seaside inns and small hotels. Myrtle Lodge (circa 1939), Imperial Court, and Driftwood Inn (circa 1940s), constructed of stucco and tile, joined the venerable guest houses of the early twentieth century: Seaside Inn, Lafayette Manor, Myrtle Beach Yacht Club, and the luxurious Ocean Forest Hotel. Three properties remaining from the 1940s—Pleasant Inn, Chesterfield Inn, and Rainbow Court—are listed in the National Register of Historic Places.

The Myrtle Beach Pavilion (circa 1948) provided an entertainment venue for music and dancing. The adjacent Myrtle Beach Pavilion Amusement Park added to the attraction. While still a vacation destination for families, the site attracted teenagers, who began to come unchaperoned toward the end of the 1950s, and the tradition of a Myrtle Beach spring break was born. *Shag, the Movie,* starring Phoebe Cates and Bridget Fonda, is a coming-of-age movie set in Myrtle Beach during summer 1963. Filmed in 1989, *Shag* utilized Ocean Boulevard's streetscape south of the Pavilion, which was still evocative of the 1960s.

Myrtle Beach was a proving ground for bands that later became mainstream performers. The Platters, Temptations, Tams, and Four Tops all played at the Beach Club.

It was the music, according to Bobby Taylor, a white blues musician, that bound local whites and blacks together during the 1960s. An after-hours club, the Blue Moon, was a black club where all people who appreciated rhythm and blues could go.

"You had to be a local or know someone," remembered Taylor, "'cause they would cut you and throw you out the door if you showed any bias."[2] Blacks and whites danced together, sometimes as couples. At the Pavilion on Friday nights, however, a small percentage of the crowded dance floor would be black, and the races did not intermingle. Taylor does not remember any racial incidents at the Pavilion except when "farm boys from Georgia would come and not understand how things were." One of Taylor's fond memories was hanging out in front of the Pavilion when he was about thirteen years old and hustling tourists for quarters so that he could get inside and hear the music.

MOTELS

Myrtle Beach's popularity increased through the 1940s with additional tourist attractions, and future growth seemed unlimited. Then, in 1954, nature interfered with progress in the form of Hurricane Hazel, a category 4 storm. It struck the East Coast of the United States on October 15, causing 1.5 billion dollars in damage and killing more than one hundred Americans in addition to the more than one thousand Haitians and eighty-one Canadians. With sustained winds of 140 miles per hour, it has been described as one of "the most intense, costliest, and deadliest hurricanes in American history."[3] The total estimated damage to the entire Grand

FIG. 10.3. Riptide Motel. Photo by Joel Carter, AIA, 2003.

Strand, ranging from Pawleys Island to Cherry Grove, was 24 million dollars. In Myrtle Beach 273 houses were destroyed, representing 80 percent of the waterfront development.[4]

A recent county history has described the region after Hazel: "many of the small tourist operators did not have the resources to rebuild, but those who did have the financial strength were able to buy land on which to put up larger hotels."[5] The Myrtle Beach Chamber of Commerce began an aggressive post-Hazel marketing campaign to let the country know that the area had recovered and was open for visitors. The chamber had five principal areas of activity to promote the expansion of Myrtle Beach's economy: travel promotion, business development, community development, public affairs, and membership affairs.

First and foremost was travel promotion. During the ten-year period following Hazel, private investment of 40 million dollars was made in lodging, business, and residential facilities. In 1963 tourism attracted a record total of 48.5 million dollars to the

FIG. 10.4. Wayfarer Motel. Photo by Joel Carter, AIA, 2003.

Myrtle Beach area. It was estimated that 37 million dollars of this figure represented vacation expenditures of visitors who stayed at local lodging accommodations.[6]

Motels became the dominant form of lodging, spreading for fifty blocks along Ocean Boulevard, from 21st Avenue North down to 29th Avenue South, although it was not until 1971 that the chamber dropped the term *motor court* from its travel guides in favor of the more modern *motel.* The motels built during this period of economic expansion were family owned and operated, ranging from a minimum of ten rooms to a maximum of one hundred. The style was Populuxe.

Thomas Hine, in his book *Populuxe,* notes that the motels built in the United States during the period between 1945 and 1963

> tended to be small, one or two story buildings of simple construction, but with signs and check in offices that generally tried to combine the image of efficient modernity with the fun of being on vacation. Many of these were torn down as the interstates were built and replaced by chain motels at the interchanges. Those that do survive tend to be in resort areas. Places like the Wildwoods, on the New Jersey Shore, and Myrtle Beach, South Carolina, still have concentrations of Populuxe motels.[7]

FIG. 10.5. The Diplomat Motel. Photo by author, 2002.

Other names describing the culture, architecture, and design of the Populuxe period include post–World War II, Googie, Jetsonian, Jetset, and Space Age. In *Main Street to Miracle Mile,* Chester H. Liebs used the phrase "Exaggerated Modern" to describe the commercial architecture of the 1950s and 1960s, where the architect adhered to modernism's functionality while exaggerating the building's structural elements. "In this form of overstated functionalism, concrete, glass, plywood, steel, plastics, and other industrial produced building materials were crafted into displays of technological exhibitionism, designed for effect more than for structural requirements."[8]

Morris Lapidus, designer of Miami Beach's Fountainbleu Hotel and Eden Roc—hotels defined by curved lines, squiggles, and woggles—was at the forefront of the Populuxe architectural movement. While guests loved his buildings and their embellishments, critics were not so enthusiastic. To them, he responded, "It's not done tongue in cheek. It's done because people need something to give them a lift. It's the crazy hat for women, the bright tie for men. Nobody can say that in order to be modern our buildings must be cold, rectangular, barren, simple planes."[9]

Myrtle Beach's motels needed embellishment. The typical concrete block structure was two to four stories with a flat roof projecting over exterior hallways. Concrete block construction dominated Myrtle Beach motels for logical reasons. It was cheaper than frame construction, went up quickly, did not require a skilled labor force, and was thought to be able to withstand hurricanes, an important consideration after Hurricane Hazel. And, unlike metal, concrete would not succumb to rust and corrosion in the salt air. Rectangular L and U shapes were most common. Rooms were small with none of the amenities taken for granted today; they were designed for sleeping or for changing clothes between beach and entertainment activities. The lobbies were most commonly carved out of first-floor rooms, faced the parking lot, and consisted of check-in counters and brochure racks. Unlike hotel lobbies with large, furnished spaces, the motel lobby was not designed for lingering, but for efficiency.

FIG. 10.6. Glazed blue brick provides a contrast to the white screen wall at the Fountainbleu Inn. Photo by the author, 2002.

Signs were fairly subdued: no neon, no flamingos. Names were evocative of the beach—Sea Gypsy, Sea Cove, and Palm Crest—or luxury, such as the Diplomat, Fountainbleau, and Continental. The most distinctive design feature of Myrtle Beach's motels was the concrete screen wall, used as both a decorative and practical element (fig. 10.6).

Between 1960 and 1964 twenty-two motels in this style were listed in the city directory, and another eighteen from 1965 through 1968. The design proved so enduring that many motels built through the late 1970s were in the same style, more than a decade after the end of the Populuxe era.

SCREEN WALL

The National Concrete Masonry Association defines screen walls as "those wall structures constructed from concrete shapes having over twenty-five percent open areas."[10] Screen walls can normally carry their own weight up to twenty feet and are not load bearing. Joint-strengthening rebar (or reinforcing bar), used every other course with pilasters built every ten to twelve feet, helped stabilize and give lateral support to large walls. Edward Durell Stone's U.S. embassy in New Delhi (circa 1954) "thrust screen block into the international spotlight,"[11] but as stated in *Palm Springs Weekend,* it was "vernacular architecture that took the sculptural concrete block the furthest, introducing elegant fleur-de-lis patterns, varied circles as random as bubbles in a tumbler of seltzer, and op-art circles and squares. Easy to design, easy to cast, a concrete block screen could add functional ornament to any apartment, shop or home."[12] Throughout Myrtle Beach ornamental block can be found in domestic and commercial architecture of the Populuxe period (fig. 10.7).

According to the North Carolina Concrete Masonry Association, concrete screen wall was "the least costly of all modern, high quality building materials."[13] Manufactures of screen block claimed that buildings using this technique were ten degrees cooler in the summer, an important feature in the South. The Adams Concrete Products Company of Kinston, North Carolina, manufactured twelve patterns in four colors in its Mar-Vail line, advertising that "these twelve designs offer an almost unlimited versatility that ranges from the classically simple to the intricately detailed and sophisticated."[14] Marketed to discriminating architects, Mar-Vail screen wall units were designed to "allow measured sunlight in pleasing dosages."[15]

A 1960s brochure for Q Block, aimed at homeowners and published by the National Concrete Masonry Association, called concrete screen wall the "modern quintessence of outdoor living. It is the least costly of *all* modern, high quality outdoor building materials. Like the artist with his paints, it is *how* you creatively employ various sizes, shapes and textures of Q Block in our lot 'frame' that deco-

FIG. 10.7. The flat roof, exterior hallways, and exposed rafters of the Virginian are typical of Myrtle Beach motels. Photo by the author, 2002.

ratively delivers the drama and distinction you desire."[16] The trade association also promoted the product's durability: "They can never rot or rust. Time takes no toll either, for Q Block becomes even more beautiful as time goes on."[17]

Not everyone, however, agreed with that assessment. Anthony Rubano, in his paper "The Grille Is Gone: The Rise and Fall of Screen Block," noted that "by the late 1960s, the masonry screen wall that previously knew 'no limitation—either geographic or artistic' began to show its age. No new block design can sever the link between screen block and the Populuxe era."[18] That in itself is not a bad thing. Dating a building by its architectural elements helps to place it in the context of the period it was constructed. More troubling was Hine's assessment that the Populuxe motels "probably won't last much longer because these buildings, which were cheap to begin with, are showing their age badly and will probably be renovated in a fashion that appears more up-to-date."[19] More than fifteen years after that was written, Myrtle Beach's remaining Populuxe motels are, for the most part, well maintained, well preserved, and still popular among repeat guests and those who do not need the amenities of a hotel.

FIG. 10.8. The Diplomat Motel (1962) is one of the best-preserved Populuxe buildings in Myrtle Beach. Photo by the author, 2002.

FIG. 10.9. The Victory Motel shares many of the later Populuxe features. Photo by the author, 1999.

CONCLUSION

The shag's popularity proved so enduring that in 1984, the South Carolina General Assembly adopted it as the official state dance. Act No. 328 called this dance "one of the greatest developments of terpsichorean culture."[20] Act No. 15 in 2001 designated beach music as the official popular music, recognizing that "beach music has contributed tremendously to the enjoyment of our citizens and has become synonymous with the Shag."[21] Each year a total of more than twelve thousand Society of Stranders members, shaggers of all ages, arrive on the Grand Strand for the Spring Safari or the Fall Migration to celebrate both the dance and the music.

A logical conclusion would be that, since beach music and the shag are officially recognized for their contribution to the state's culture and history, the architecture of this period would be similarly acknowledged and celebrated. To date that has not been the case. Only forty Populuxe motels remain in Myrtle Beach. The loss of these structures is unfortunately part of a national trend. With respect to roadside culture, the National Park Service found that "federal tax laws, competition within industry, changing transportation routes, and shifts in consumer tastes have jeopardized many early motel or motor court complexes, shopping centers, and other roadside buildings. Their rate of survival with integrity from the post–World War II era is very low."[22]

Rampant growth does not create an environment conducive to preservation. The increase in tourism to Myrtle Beach between 1973 and 1998 was astronomical. The number of visitors more than doubled, from 3,408,000 in 1973 to 7,700,000 in 1998. Tourism expenditures, including lodging, meals, and entertainment, jumped from 2.2 million dollars to 2.1 billion dollars over that same period.[23] The increase in visitors led to a need for more lodging accommodations. And, since oceanfront lots in Myrtle Beach are a limited commodity, high land value in an area zoned for high density created a threat to the existing structures. With the land more valuable than the structures occupying it, motels quickly gave way to high-rise hotels. In an essay titled "Historic Preservation: The Process and the Actors," Robert Stipe wrote that in order for preservation efforts to succeed, the bottom line of property owners must be considered. "Unlike many other countries in which land tends to be regarded as a scarce resource to be treated with respect, Americans always tended to view real estate as a marketable commodity whose principle purpose is to provide capital gains or income to its temporary owner."[24]

In Myrtle Beach economic development and not historic preservation has driven the development process for decades. The effects of economic prosperity and increased tourism have not been all positive. In the race to build new attractions to accommodate an increasingly volatile tourist base, Myrtle Beach has sacrificed most of its history. None of the Populuxe motels was designed by noted architects.

Indeed, no one is making the argument that these are great works of architecture. But they provide a tangible link between the beginning and end of the twentieth century in Myrtle Beach's history as it relates to tourism. They provided, and still provide, opportunity for small business entrepreneurs. They also fill a market niche not served by new high-rise hotels: personal service at affordable rates.

For now it is still possible for a visitor to enjoy Myrtle Beach as it was more than four decades ago. The daring can grab a stool at the lunch counter in Peach's Corner, a local landmark since 1937; order a chili dog all the way; and then ride the Swamp Fox, a circa-1961 wooden rollercoaster at the Family Kingdom Amusement Park. The not-so-daring can relax by the artfully screened pool at a Populuxe motel and tune to 94.9 FM—beach music and rhythm and blues all the time.

NOTES

This essay was reprinted from *SCA Journal* 21 (Spring 2003): 4–9. Used with permission.

1. Lena Lencek and Gideon Bosker, *The Beach: The History of Paradise on Earth* (New York: Penguin Books, 1999), 244–45.
2. Bobby Taylor, interview by author, August 21, 2000.
3. Storm Signals Online, http://www.stormsignals.com/hazel.html (accessed February 19, 2002).
4. Weather Catalog, http://i4weather.net/hazel.html (accessed February 20, 2002).
5. Catherine Lewis, *Horry County, South Carolina, 1730–1993* (Columbia: U of South Carolina P, 1998), 78.
6. Greater Myrtle Beach Area Chamber of Commerce, *The Role of Advertising and Travel Promotion in the Economic Development of Myrtle Beach, South Carolina: A Decade of Progress, 1954–1964* (Myrtle Beach: Greater Myrtle Beach Area Chamber of Commerce, 1965), 4, 10.
7. Thomas Hine, *Populuxe* (New York: Knopf, 1986), 157.
8. Chester H. Liebs, *Main Street to Miracle Mile* (Boston: Little, Brown, 1985), 63.
9. Morris Lapidus, *Too Much Is Never Enough* (New York: Rizzoli, 1996), 196.
10. Anthony Rubano, "The Grille Is Gone: The Rise and Fall of Screen Block," in *Preserving the Recent Past 2*, ed. Deborah Slaton and William Foulkes (Washington, DC: Historic Preservation Education, 2000), 3–90.
11. Ibid.
12. Alan Hess and Andrew Danish, *Palm Springs Weekend: The Architecture and Design of a Midcentury Oasis* (San Francisco: Chronicle Books, 2001), 108.
13. Adams Concrete Products Company promotional brochure, n.d.
14. Adams Concrete Products Company, Mar-Vail Screen Wall Units sales brochure, early 1960s.
15. Ibid.
16. *Q Block: Modern Quintessence of Outdoor Living*, vol. 20, no. 3, National Concrete Masonry Association, Arlington, Va., undated brochure.

17. Ibid.

18. Rubano, "The Grille Is Gone," 3–95.

19. Hine, *Populuxe,* 157.

20. South Carolina Information Highway, www.lpitr.state.sc.us/studentpage/shag.html (accessed August 28, 2003).

22. South Carolina Information Highway, http://www.liptr.state.sc.us/studentpage/pop music.html (accessed August 28, 2003.

22. Slaton and Foulks, *Preserving the Recent Past 2,* 1.

23. South Carolina Department of Parks, Recreation, and Tourism, *Direct Impact of Visitor Spending and New Tourism Development in South Carolina by County: 1998* (Columbia: State of South Carolina, 1999).

24. Robert Stipe, "Historic Preservation: The Process and the Actors," in *The American Mosaic: Preserving a Nation's Heritage,* Robert E. Stipe and Antoinette J. Lee, eds. (Washington, D.C.: US/ICOMOS, 1987), 5.

CHAPTER 11

CAVING AND CLOGGING: KEEPIN' COOL IN TENNESSEE CAVES

RUTH NICHOLS KEENOY AND ROBBIE D. JONES

The dark mysterious nature of caves has intrigued humans for centuries, leading to explorations from a sense of adventure as well as for practical purposes. For example, American Indians camped in Kentucky's Mammoth Cave and European American settlers used it as a saltpeter mine and as a tuberculosis hospital. By the twentieth century caves were used more recreationally, even as respectable sites for weddings and community dances or as illicit speakeasies and locations for illegal moonshine stills. With few exceptions, caves did not become popular tourist attractions until the construction of good local and interstate roads during the 1920s. These roads, such as the Dixie Highway, provided stable, year-round expanded access for residents and tourists to visit caves, particularly in the hills of Tennessee, where caves are numerous. Roads were the keys to the life of these facilities as tourist attractions, and once bypassed by the interstate system of the 1950–60s, many of these isolated caves closed. Most caves that remained open did so only seasonally, usually for cave tours, or because they were easily accessible from the new interstates. Today many Tennessee caves remain popular tourist attractions, and several are listed on the National Register of Historic Places.

During steamy summer vacations on the road, many automobile travelers stop at roadside "show caves" and take a break from the heat by going underground. Located in limestone-rich regions across the United States, subterranean show caves not only provide overheated and road-weary travelers with an opportunity to cool off, but also offer captivating tours and breathtaking scenery. Usually located near highways with heavy tourist traffic, commercial show caves are among the more unusual examples of the roadside tourism industry. Their success as attractions is directly associated with the development of good roads and early "interstate" highways, such as the Dixie Highway, which provided vehicular access to previously rural or isolated areas and also provided commercial cave owners with an advertising venue that would attract more visitors.

In addition to their use as show caves, the cool underground surroundings created an opportune setting for other forms of commercial activities. This was especially true in the South during the hot summer months and particularly in Tennessee, which has approximately seven thousand caves, more than have been documented in any other state. Southerners have used caves for a plethora of activities, including farming mushrooms and canning tomatoes, preaching sermons and hosting weddings, and housing sick babies and tuberculosis patients.[1]

Some of Tennessee's caves were used as unlicensed saloons where moonshine or whiskey was served and as havens for illicit activities such as gambling on card games and cockfights. Southerners claim that malicious spirits haunt some caves, such as Tennessee's famed Bell Witch Cave, and sell tickets for supernatural tours. Perhaps it is the commercial dance cave that serves the most extraordinary purpose. For more than a century, Tennesseans have gone underground to clog and dance to eclectic musical arrangements performed by big bands, orchestras, country fiddlers, and members of the Grand Ole Opry.[2]

The commercial development of southern caves began as early as 1816, with tours of Kentucky's Mammoth Cave, one of the oldest U.S. tourist attractions. Automobile tourists flocked to Mammoth Cave during the early twentieth century, which led to its eventual designation as a national park in 1941. The immense success of Mammoth Cave as a roadside tourist attraction spurred the development of numerous commercial show caves throughout the region. Tennessee's tribute to Mammoth Cave's popularity was the "New Mammoth Cave," the largest cave in the Cumberland Mountain region, which opened in 1921 near the route of the Dixie Highway at Jellico in East Tennessee, near the Kentucky state line.[3]

Additional commercial show caves located on or near the Dixie Highway include Cudjo's Cave near Cumberland Gap where Tennessee, Virginia, and Kentucky meet. This extensive cave system is located near the legendary Wilderness Trail, first used by American Indians, and later by white pioneers such as Daniel Boone, to

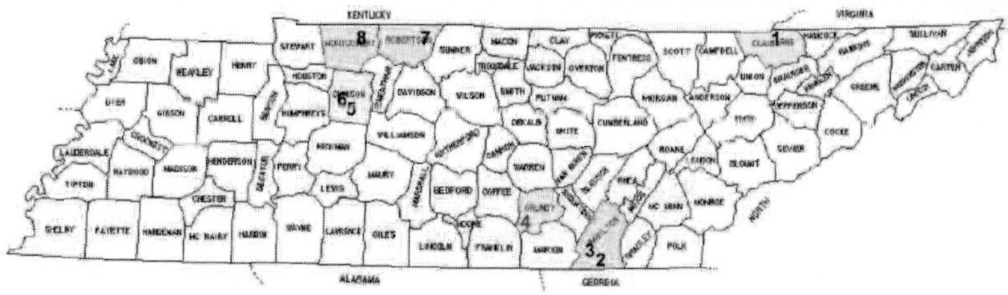

1. Cudjo's Cave near Cumberland Gap
2. Ruby Falls on Lookout Mountain near Chattanooga
3. Tennessee Caverns on Raccoon Mountain near Chattanpoga
4. Wonder Cave near Monteagle
5. Jewel Cave near Dickson
6. Ruskin Cave near Dickson
7. Yates Cave near Cross Plains
8. Dunbar Cave near Clarksville

FIG. 11.1. Map of tourist caves in Tennessee. Courtesy of Martha Carver.

cross the rugged Appalachian Mountains. This route eventually became the Dixie Highway and later U.S. 25E. As was first documented in 1750, early settlers mined this five-tier cave for saltpeter, while outlaws and runaway slaves used it as a secure hiding place. The cave was originally called Soldiers Cave because soldiers used one of the caves as a military hospital during the American Civil War and inscribed their names on the walls. As early as the 1890s, entrepreneurs operated tours of the cave (fig. 11.1).

Lincoln Memorial University, a small private college located in nearby Harrogate, Tennessee, purchased the entire cave system in 1920, and contracted with concessionaires to provide commercial cave tours as a fundraising venture. Around 1934 the university improved the cave with electric lighting, walkways and bridges, a new entrance, exit doors, stairs, and connecting tunnels. At that time the university renamed the improved site Cudjo's Cave, after a popular Civil War novel of the same name written by John T. Trowbridge. The university constructed a small rustic log building that housed a store, ticket office, and souvenir stand along the Dixie Highway.[4]

Famous for its "fantastic and bewitching beauty," Cudjo's Cave was also popular as an underground dance hall. Promotional material boasted that the cave held a subterranean "cathedral" and the world's biggest stalagmite, measuring sixty-five feet in height and thirty-five feet in circumference and appropriately called the "Pillar of Hercules."[5]

Although Cudjo's Cave was incorporated as part of the Cumberland Gap National Historical Park in 1947, the university continued to operate it as a roadside tourist attraction until 1992, when the National Park Service (NPS) purchased the cave. In 1996 U.S. 25E was rerouted through a nearby tunnel beneath Cumberland

Mountain, and the old Dixie Highway was eradicated. In an effort to restore the cave to a "more natural state" and to rebuild the Wilderness Trail nearby, the NPS recently demolished all twentieth-century features of Cudjo's Cave.[6]

Perhaps Tennessee's most successful commercial show caves along the Dixie Highway were Chattanooga's Lookout Mountain Caverns, which feature the renowned 145-foot underground waterfall known as Ruby Falls. Located on Lookout Mountain near Chattanooga, the site consists of two caves, upper and lower, that opened to the public in 1929 and 1930, respectively. The lower cavern, no longer accessible, was used extensively by American Indians and later by Union and Confederate soldiers during the Civil War. Even Andrew Jackson visited, leaving his signature inscribed on the cave's wall. The upper cave, located approximately 260 feet underground, remained one of Chattanooga's best-kept secrets until 1928, when the Lookout Mountain Cave Company accidentally discovered it during blasting for development of a 400-foot elevator shaft.[7]

Commercial development on Lookout Mountain began during the mid–nineteenth century, when an early toll road and the railroad opened the area to tourism. In 1857 the Lookout Mountain Hotel began hosting summer tourists, and by 1860 the mountain area averaged seventy-five thousand tourists annually. In 1886 Lookout Mountain gained one of its unique tourist attractions when the Lookout Point Incline Company began operating a passenger railway from St. Elmo, a streetcar suburb south of Chattanooga, to the top of the mountain. This tradition continues today with Lookout Mountain Incline Railway, established in 1895. Touted as "America's Most Amazing Mile," the railway, known locally as simply "The Incline," provides passengers with a breathtaking view from its steeply graded rail bed, and it remains one of Tennessee's most popular roadside attractions.[8]

Some people credit Leo Lambert as the individual responsible for the development of Lookout Mountain Caverns. A native of Indiana, Lambert moved to Chattanooga in 1916 to marry his childhood sweetheart, Ruby Losey. Shortly after his arrival to Chattanooga, Lambert learned about the lower cave from a Southern Railway employee. Intrigued by the idea of opening the cave commercially and based on the success of nearby Lookout Mountain's Rock City Gardens, which opened to automobile tourists in 1925, Lambert returned to Gary, Indiana, where he gained financial backing through local investors. Several Chattanooga businessmen provided additional funding. By the time Lambert established the "Lookout Mountain Cave Company" in 1928, he had acquired financial backing of 250,000 dollars.[9]

When the upper cave opened in 1929, promoters advertised the cave as featuring the world's "largest known underground waterfall." The spectacular cascade Ruby Falls was appropriately named for Lambert's wife, who accompanied her husband on his second visit to the cave. The upper cave also featured the "longest ele-

FIG. 11.2. Constructed in 1929, Cavern Castle remains in use today, housing a gift shop, with ticket sales for and elevator access to Ruby Falls. Circa-1935 postcard, from the collection of Ruth Nichols Keenoy.

vator ride south of the Mason-Dixon Line." Amenities included telephones, wooden bridges, and electric lights that consisted of bare light bulbs strung on wires throughout the cave. Cave tours began every twenty minutes, and an early promotional brochure boasted that there was "never a dull moment for the tourist." Visitors entered the caves via Cavern Castle, a three-story building constructed from limestone removed during construction of the elevator (fig. 11.2). Depicted as the "most beautiful cave entrance in the world," the castle featured elevator access, a gift shop, a restaurant, administrative offices, and space that could be rented for "meetings, parties, and dances." An adjacent parking area accommodated up to two hundred vehicles. Opening festivities included a week of live music and dancing with a grand opening ceremony that included music by the Frank Worthip Orchestra. In 1931 Babe Ruth and the New York Yankees toured Ruby Falls.[10]

Unfortunately, Lambert's timing coincided with the Great Depression. Declining tourism forced the development company into bankruptcy by 1932, at which time the caves were sold to Claude Brown of Chattanooga for a mere twenty-five thousand dollars. Brown continued cave tours throughout 1930s, promoting dances and live music held at the Cavern Castle in hopes of stimulating tourism. In 1935 the lower cave was closed because the railroad had caused increasing soot deposits. The upper cave was sold in 1947 and again in 1949, at which time the Lookout Advertising Company gained partial ownership and aggressively promoted the cave as a roadside tourist attraction. With the national rise in tourism after the war, in addition to the site's location near the well-traveled Dixie Highway and a massive advertising campaign, Lookout Mountain Caverns and Ruby Falls finally achieved Lambert's intended financial and commercial success. Today Ruby Falls, which underwent eight hundred thousand dollars in improvements in 2003, remains one of Tennessee's most popular tourist attractions, especially with families and school groups.[11]

Before Lambert's bankruptcy in 1932, he also developed Tennessee Caverns as a commercial show cave site. Currently known as Raccoon Mountain Caverns, it is located seven miles west of Chattanooga along U.S. 41, once called the "new Dixie shortcut." Tennessee Caverns opened on July 29, 1931, under the guidance of Lambert. Promoters identified the site as "another important step in the building of a cave region in Chattanooga and the vicinity rivaling any system in the country." A "Kohler light plant" provided electric lighting throughout the cave, and the cave's numerous limestone formations were emphasized as the primary attraction. On entrance to the cave via an artificial opening with a crystal onyx gate, visitors found themselves in a large open "room" filled with "delicately colored" stalactites and stalagmites.[12]

As construction continued on the entry pavilion in 1931, Lambert and his investors planned the development of an automobile tourist camp and a nearby spring, to be called the "Raccoon Mountain Recreational Park." Again Lambert's venture landed on the brink of the Great Depression, which forced him into selling the entire site to Claude Brown, presumably in 1932.[13]

In 1952 M. L. C. Smith purchased Tennessee Caverns and changed the name of the site to Crystal City Caverns. Throughout Smith's years of ownership, guides directed visitors to turn their attention toward formations throughout the cave that were promoted as having the uncanny likeness of various biblical characters, such as Daniel in the lions' den and Adam and Eve in the Garden of Eden. In the 1960s a tram was constructed that provided "sky rides" from the cave to the top of Raccoon Mountain. It was also during this period that the cave managers promoted a summer seasonal American Indian "village." This village sponsored "all-Indian" performances featuring "Indian songs and dances."[14]

Tennessee Caverns continued to change hands throughout the 1970s and 1980s. In 1995 Steve and Bob Perlaky purchased the site. Today visitors enter the cave through the gift shop, which is cooled naturally during the summer months by the cave's "blowing" air. A large campsite located adjacent to the cave features a recreational vehicle park, a swimming pool, and hiking trails. Although no longer associated directly with the cave, the sky tram continues to operate.

Located along U.S. 41 in Grundy County, four miles northwest of the resort of Monteagle, is Wonder Cave, considered by many cave enthusiasts to be one of the state's preeminent show caves. Wonder Cave operated as a commercial cave for more than a century. Three Vanderbilt University students, Robert A. Nelson, Melville Anderson, and Will Fitzgerald, discovered the cave in 1897. A year later owner R. M. Payne enlarged the cave's natural entrance and opened the site to the public. The Mystic River runs through the cave, and early visitors entered via flat-bottom boats. Payne also constructed trails throughout the cave and provided lighting using a gas generator. Later attempts to light the cave electrically proved unsuccessful because wintertime flooding repeatedly shorted out the electrical system. In most recent years, guided tours relied on lighting provided by handheld Coleman lanterns.[15]

During the early 1900s the site's subsequent owner J. J. Raulston constructed a two-story log dwelling near the cave's entrance. This impressive building held a gift shop and ticket office and served as Raulston's personal residence; it later functioned as a bed-and-breakfast. A stone entry facade leading to the cave (fig. 11.3), also constructed by Raulston around 1900, remains intact, as does early-twentieth-century religious statuary erected by subsequent owners. Guided boat tours along the Mystic River, which originally provided sole entry to the cave, were discontinued in 1917, following Raulston's improvements to the site.[16]

For many years Wonder Cave consistently attracted more tourists than Ruby Falls or Raccoon Mountain did. This was in large part because the site was adjacent to U.S. 41. For travelers heading south, the cave provided the state's first major roadside tourist attraction. Many vacationers caught their first glimpse of the Appalachian Mountains here and preferred stopping at Wonder Cave rather than continuing south to the caves near Chattanooga. Cave records from 1961 document that more than forty thousand tourists visited the cave in that year alone. The construction of Interstate 24 approximately two miles west of the cave drastically altered these statistics. When the interstate opened in 1962, Wonder Cave's visitation decreased by 90 percent, marking the sudden demise of the site's status as Tennessee's most popular show cave. More than two million visitors were reported to have toured Wonder Cave between 1897 and 1987.[17]

Not all show caves in Tennessee depended on the proximity to major highways that carried large numbers of tourists for their commercial success. Located in rural Dickson County on State Route 46, Jewel Cave is perhaps Middle Tennessee's most spectacular show cave. Although it is one of the smallest commercial caves in Tennessee, many visitors consider it the most beautiful. In 1940 a Work Projects Administration writer described Jewel Cave as a "subterranean fairyland" because of its numerous sparkling crystals. Discovered in 1885 by Thomas Rogers and his daughters while exploring a spring on his farm, the cave's extravagant and colorful rock formations caused a mild sensation with the locals. When W. H. "Pop" Lawson

FIG. 11.3. Raulston's limestone entry facade to Wonder Cave remains unaltered since original construction circa 1900. Historic postcard, date unknown, from the collection of Ruth Nichols Keenoy.

opened the cave for commercial tours in 1924, he constructed a rustic ticket office that held a gift shop, restrooms, a refreshment stand, and a new entrance into the cave.[18]

Lawson installed electric lighting in the cave to display the thousands of stalactite and stalagmite formations that glistened mysteriously with "magical charm." Most of the rock formations have been tinted by mineral water slowly seeping through them for thousands of years, resulting in a rainbow of pastel shades of orange, blue, purple, and red. It is this "resplendent beauty" and the "mysterious formations" that led to the cave's designation as "Jewel Cave."[19]

Lawson gradually improved the interior of the cave with concrete stairs and concrete handrails cast to simulate wood. In 1930 Lawson discovered an adjacent cave, which he added to the tour a few years later after hand-digging a rock-lined connecting tunnel. Still open for seasonal tours, little has changed about Jewel Cave since the 1930s. It remains an awe-inspiring place to keep cool from the hot summer sun (fig. 11.4).

FIG. 11.4. Built in the 1920s, the rustic ticket office at Jewel Cave is an exceptional example of Tennessee's roadside architecture. Historic postcard, date unknown, from the collection of Robbie Jones.

Located less than half a mile south of Jewel Cave is Ruskin Cave, one of Tennessee's most documented historic caves. In 1896 a group of socialists purchased the 384-acre farm containing this cave for use as a utopian cooperative village. Established two years prior by socialist leader Julius Wayland, the colony was named "Ruskin" in honor of the celebrated English social reformer John Ruskin. Following their purchase of the cave site, the 250 colonists quickly constructed 75 buildings centered on an imposing three-story "Commonwealth Building" that housed a community dining hall and a 700-seat auditorium where plays and concerts were held. This building also featured a 1,500-volume library, bookstore, barbershop, nursery, overnight lodging facilities, stockrooms, and a printing office that published the popular socialist newspaper *The Coming Nation.*[20]

The colony itself centered its activities on Ruskin Cave, the "true heart" of the utopian settlement. Featuring a large opening on the side of a vertical limestone bluff and an underground stream, this enormous cave was used for preparing and storing canned foods, such as blackberries and tomatoes, which were in turn sold at the general store and through mail-order catalogs. The colonists also raised money by charging admission to a grand Independence Day barbecue at the cave where the Ruskin Orchestra and the Ruskin Mandolin Club played for square and circle dances, as well as for entertainments featuring waltzes, two-steps, polkas, and schottisches (round dances similar to polkas).[21]

Comprising secular intellectuals and educated romantics, as well as anarchists and deadbeats, the utopian community suffered from contentious internal debates over politics, religion, and sex. Some locals even accused the colonists of practicing "free love" in the darkened recesses of Ruskin Cave. These disputes polarized the colony and ultimately led to its collapse in 1899. The following summer the colonists held the final Independence Day celebration at Ruskin Cave and auctioned all their belongings before most left for a new Ruskin colony located in Georgia.[22]

Between 1904 and 1925 the old Ruskin Colony buildings hosted a series of private religious-based schools, including the Ruskin Cave College and Conservatory of Music (fig. 11.5). During these years the cave was used for sermons and picnics. Subsequent owners developed the property into a summer retreat known as Ruskin Cave Resort and constructed a "trailer camp, a large 75 room hotel, tourist cottages, and camping space." In 1938 Ruskin Cave was used as the "world's largest underground ballroom" when a concrete dance floor and bleachers were added. Reportedly more than 13,300 people attended the cave's opening dance, which featured Nashville's Francis Craig's Orchestra; Craig was a popular bandleader, pianist, and songwriter as well as a regular on Nashville's radio station WSM, known for its far-reaching broadcast of the Grand Ole Opry.[23]

Ruskin Cave's traditional Saturday night dances soon featured a unique type of fiddling known as "country and bluegrass." Billed as "Dixie's Coolest Play Ground," Ruskin Cave Resort featured an in-ground swimming pool and a stone concession stand with restrooms. With the introduction of local "hillbilly" music, Ruskin Cave became well known for hosting future Grand Ole Opry performers. Although the Ruskin Resort closed in 1941, Joe Gibson resurrected the "Saturday Night Hillbilly Hoe-down" in 1959. Country and bluegrass music entertainers dominated the Ruskin "cave dances," but other musical styles, such as "old-time" square dances and gospel, were also important throughout the 1960s and 1970s.[24]

FIG. 11.5. Ruskin Cave is one of the state's most historic dance caves, shown here during its use as Ruskin Cave College in the 1910s. Historic postcard, date unknown, from the collection of Ridley Wills.

Country music "outlaw" David Allan Coe owned and operated the Ruskin Cave Resort throughout the 1980s. Most famous for writing *Take This Job and Shove It,* the songwriter, a former convict, wrote many of his biggest hits while living at Ruskin, where he continued the tradition of rowdy Fourth of July concerts until he departed in 1988. Coe left his mark at Ruskin, but his unique stone house was left unfinished. Only a handful of the original Ruskin buildings remain standing, including the historic Commonwealth Building, although visitors can still find many remnants of the cave's rich history.[25]

Like Ruskin Cave, many rural Tennessee caves operated as summertime recreational attractions for local populations, including Yates Cave near Cross Plains in Robertson County. Located on the "Triple Cave Farm" near State Route 25, this relatively large cave features a small entrance at the bottom of a thirty-foot limestone bluff that arches above an underground stream. Walton Yates purchased the cave in 1905 and promoted it for various social uses. One cave chamber was used as a "men's club" where locals drank whiskey and gambled on card games during Prohibition. Yates also promoted the cool grounds just outside the cave's entrance as the perfect site for family picnics, church reunions, weddings, and Sunday outings.[26]

Around 1930 Yates constructed an entrance and storage room at the side of the cave. The room housed a refreshment stand where he sold sandwiches, soft drinks, and homemade ice cream. Yates dammed a stream to create a swimming hole outside the entrance and began selling tickets for guided boat tours into the cave's thirty foot by thirty foot "rock chamber." He installed electric lighting and paneled walls inside the cave, as well as a concrete floor where local residents held dances and picnics. After Walton Yates died in 1937, his son Sowell continued operating the cave until the 1940s. Today the old refreshment stand remains a reminder of this cave's remarkable past.[27]

Another distinctive commercial cave in Tennessee is Montgomery County's Dunbar Cave, located two miles south of Clarksville. Documentation of the cave's discovery by white settlers dates to 1790, when Isaac Rowe Peterson of North Carolina happened on the cave. By the time Peterson returned with his family in 1791, Thomas Dunbar, for whom the cave was eventually named, had moved into the cave and constructed a fortified blockhouse near the cave's entrance.[28]

The breadth of historic activities associated with Dunbar Cave throughout the nineteenth and early twentieth centuries is vast. Local residents mined the cave for saltpeter used in the production of gunpowder during both the Mexican War and the American Civil War. The cave was also used briefly as both Confederate and Union shelters during the Civil War, for meetings held by the Ku Klux Klan, as the hide-out for a local counterfeit production, and to house illegal moonshine stills.[29]

Although Dunbar Cave was not recognized as a commercial establishment until after 1900, the site hosted numerous recreational activities before that time. Its use as a dance cave began before 1860, when the site housed a saloon and dance hall. Dances continued between 1870 and 1895, when "bran" dancing, an event in which bran was scattered across the dance floor, became popular.[30]

Shortly after the Civil War ended, James A. Tate erected the Idaho Springs Hotel near Dunbar Cave, replacing private resort cabins constructed in 1858. Train service provided transportation to the new hotel and the cave. In 1884 Col. A. G. Goodlet and H. C. Merritt purchased the site and enlarged the hotel. Advertisements at that time not only boasted of the hotel's new amenities, which included telephones and livery service, but also of the site's mosquito-free ambiance—an enticement that no doubt lured many hot, itchy tourists during the steamy southern summers.[31]

In 1931 the Dunbar Cave and Idaho Springs Corporation again renovated the cave and hotel property in hopes of developing a more profitable tourist site. The corporation constructed a dam near the cave's entrance and created a fifteen-acre stocked lake for fishing and boating. A bridge was constructed across the lake to provide access to hiking trails. Improvements also included construction of an in-ground swimming pool, tennis courts, restrooms, and a handsome Colonial Revival–style bathhouse. The corporation erected a stone concession stand and a concrete stage just outside the cave's entrance beneath an overhanging cliff, as well as an elevated terrace with railings overlooking the lower arcade that featured a concrete dance floor[32] (fig. 11.6).

Early-twentieth-century musical performances at Dunbar included big-band and country-music entertainers. Throughout the 1930s Dunbar Cave hosted dances led

FIG. 11.6. View of Dunbar Cave featuring the cave entrance, concrete dance floor, concession area, and restrooms. Historic postcard, date unknown, from the collection of Ruth Nichols Keenoy.

by acclaimed bandleaders such as Artie Shaw, Guy Lombardo, Tommy Dorsey, and Glenn Miller; the cave was a regular gig for entertainers working the "Louisville to Memphis" circuit.[33]

Despite the cave's strong association with big band music, it was country music that dominated Dunbar's regional association as a dance cave. In 1945 Clarksville mayor William Kleeman purchased the four-hundred-acre resort area. Two years later Roy Acuff, known as the "King of Country Music," purchased the cave from Kleeman, in hopes of establishing the South's first "country music park." Acuff put his band members, the Smoky Mountain Boys, to work cleaning up the "tangles of weeds and dilapidated buildings," to create what would become Roy Acuff's Dunbar Cave. Changes to the site during Acuff's ownership included demolition of the historic Idaho Springs Hotel, construction of new tourist cabins, and the addition of an eighteen-hole golf course. Acuff charged a minor admission fee and began hosting Tuesday and Friday night square dances at the cave. Friday nights were topped off with a late show featuring Roy Acuff (appropriately nicknamed "Deacon of Dunbar Cave") and the Smoky Mountain Boys that was broadcast live via Nashville's WSM radio station. The band performed in Nashville for the Grand Ole Opry's "Friday Night Frolic" between 8:00 and 8:30 P.M., then made the fifty-mile drive to the cave immediately after the show to begin the 11:30 to midnight square dance. On Friday nights when Roy and the boys were "on the road," Acuff arranged for other Opry performers to headline the weekend jamboree.[34]

During the mid-1950s the cave began to lose its popular appeal, as air-conditioning and television became staples of modern southern life. Acuff sold the property in 1963 to McKay King, who closed the swimming pool in 1967 but kept the cave open until his death in 1971. The city of Clarksville obtained the site's golf course in 1972, and the following year the state of Tennessee purchased the remainder of the property for use as a 110-acre public recreational park. At that time the swimming pool was filled in, and the bathhouse was renovated as the park's visitor center. Despite the recent changes to the site, including the removal of the cave's electric lighting system, numerous reminders of Dunbar's dance cave heyday remain.[33]

EPILOGUE

The construction of modern interstate highway systems during the 1960s and 1970s led to the closure of many of Tennessee's historic roadside commercial caves. Despite this trend, however, a new breed of commercial show caves developed during this era. Many of these caves continue to use persuasive gimmicks to entice tourists off the interstate, including "God of the Mountain," an underground light and music show at Cumberland Caverns near McMinnville in Middle Tennessee, and glass-bottomed electric-boat rides at the subterranean Lost Sea, a 4.5-acre underground lake near Sweetwater in East Tennessee.

Near Gatlinburg, Forbidden Caverns and Tuckaleechee Caverns lure automobile tourists on their way to the popular Great Smoky Mountains National Park. Some of Tennessee's new breed of show caves link their popularity to the cold war era, when they were used as bomb shelters, including Appalachian Caverns at Blountville and the aptly named Atomic Caverns near Oak Ridge. Historic show caves have adapted to new uses, such as recent circuit-club dance parties held in Indian Cave in Grainger County and Halloween "haunted caverns" at Ruby Falls and Dunbar Cave.

While underground dancing is no longer popular in Tennessee, thousands of tourists continue to visit the state's underground commercial show and dance chambers. Many stop on their way to another vacation destination, while a few come solely in search of the caves themselves. As roadside tourist attractions, Tennessee's historic commercial show caves and dance caves continue to provide their visitors not only with a cool refreshing change of climate, but also with an exciting secretive world where one can still imagine Tennessee's bygone era of "caving and clogging."

NOTES

1. Larry Matthews, Tennessee Division of Geology, telephone interview with Ruth Nichols Keenoy, August 1998; and Thomas Calhoun Barr Jr., *Caves of Tennessee* (Nashville: Tennessee Division of Geology, 1961). Matthews is author of *Descriptions of Tennessee Caves* (Nashville: Tennessee Division of Geology, 1971). More than seven thousand caves have

been documented in Tennessee, around two thousand more than in Missouri, the state with the second largest number of documented caves.

2. Ann Toplovich, "Dancing and Fiddling in the Cool of the Cave: Dance Caves in Tennessee," *Courier* 21 (June 1983): 4–5. Dance caves have also been documented in Texas and Arkansas. Barr documented a cave called "Dance Cave" in Moore County, Tennessee, in *Caves of Tennessee.*
3. Margaret M. Bridwell, *The Story of Mammoth Cave National Park, Kentucky: A Brief History* (Mammoth Cave, KY, 1952); Della Yoe, "Cumberland Mammoth Cave," Federal Writers' Project of the Work Projects Administration for the State of Tennessee, Records, 1941, Microfilm Accession Number 1067, Box 21, Points of Interest, Caves; Archives and Manuscripts Section, Tennessee State Library and Archives, Nashville, Tennessee. All FWP materials mentioned in subsequent citations are located here. The federal government approved of the development of three southern national parks in 1926, including Mammoth Cave, which achieved National Park status in July 1941. East Tennessee's New Mammoth Cave, located in Campbell County, was renamed "Cumberland Mammoth Cave" in 1928.
4. Yoe, "Cumberland Mammoth Cave," 1941; John Townsend Trowbridge, *Cudjo's Cave: A Story of the Civil War* (New York: A. L. Burt Company, [1893]). An East Tennessee Unionist originally from Pennsylvania, Trowbridge had been harassed by Confederates during the Civil War. In Trowbridge's novel Cudjo was a runaway African slave who lived in a similar cave; locals attributed that local to Soldier's Cave and renamed it "Cudjo's Cave."
5. Della Yoe, "Cudjo's Cave," Federal Writers' Project, 1941.
6. National Park Service, *Cave Management Plan for Cudjo's Cave* (Washington, DC: United States Department of the Interior, 1997); Doug Hufziger, "Cumberland Gap," *Pine Mountain Fault,* Fall 1997, 14–15. Historic properties within the boundaries of the Cumberland Gap National Historical Park were listed in the National Register of Historic Places in 1966; however, Cudjo's Cave was listed as a "non-contributing" property.
7. Federal Writers' Project, Work Projects Administration for the State of Tennessee, *The WPA Guide to Tennessee* (1939; reprint, Knoxville: U of Tennessee P, 1986), 165; Elizabeth Wordrup, "Ruby Falls," Federal Writers' Project, 1941; Della Yoe, "Lookout Mountain Caverns," Federal Writers' Project, 1941.
8. "History of the Lookout Mountain Incline Railway," in "National Historic Site" (unpublished booklet, photocopy available at the Tennessee Department of Transportation, Environmental Division, Nashville), 2–15. The Lookout Mountain Incline Railway was listed in the National Register of Historic Places in April 1973.
9. E. Raymond Evans and Vicky Karhu, Research Consultants Association for Preservation of Cultural and Natural Resources, "Lookout Mountain Caverns and Cavern Castle," Nomination Form, National Register of Historic Places Inventory, June 1985, section 8, pp. 2–3 (copy on file at the Tennessee Historical Commission, Nashville). The caverns were listed in the National Register of Historic Places in November 1985.
10. Federal Writers' Project, *The WPA Guide to Tennessee,* 265; Barr, *Caves of Tennessee,* 253; Ed Brinkley, *The History of Ruby Falls* (Chattanooga, TN.: Service Printing, 1964), 32; *Ruby Falls, Lookout Mountain Caverns, Chattanooga, Tennessee,* brochure printed by the

Chattanooga Chamber of Commerce, date unknown. Original on file at the Tennessee State Library and Archives, Nashville, in the back of Zella Armstrong, *The History of Hamilton County and Chattanooga Tennessee,* vol. 1 (Chattanooga, TN: Lookout Publishing, 1931); and Evans and Karhu, Nomination Form, section 8, p. 2, and section 1, p. 2.

11. Evans and Karhu, Nomination Form, section 8, p. 3.
12. "Cave to Open First Time Today," *Chattanooga Times,* June 28, 1931, 7 (copies of article on file at Historical Collection, Chattanooga Public Library, and Raccoon Mountain Caverns, Chattanooga, TN); Federal Writers' Project, *The WPA Guide to Tennessee,* 480.
13. Federal Writers' Project, *The WPA Guide to Tennessee,* 480; Evans and Karhu, Nomination Form, section 8, p. 2.
14. Barr, *Caves of Tennessee,* 251; Patty Perlaky, letter to Ruth Nichols Keenoy, June 17, 1998. Copies of newspaper articles enclosed with letter include "Sky Ride Is Open at Mount Aetna," *Chattanooga Times,* September 6, 1964, n.p.; "Indian and Family Arrive at Aetna for Crystal Cave Expansion Work," *Chattanooga Times,* June 4, 1966, n.p. (copies of articles on file at Historical Collection, Chattanooga Public Library, and Raccoon Mountain Caverns, Chattanooga, TN).
15. Della Yoe, "Wonder Cave," Federal Writers' Project, 1941; Federal Writers' Project, *The WPA Guide to Tennessee,* 404.
16. Robert Brandt, *Touring the Middle Tennessee Backroads* (Winston-Salem, NC: John F. Blair Publisher, 1995), 308–9; and C. Jefferson Hood, "There's Plenty of Wonderment at Wonder Cave," *Nashville Tennessean,* July 4, 1995, n.p. (photocopy on file at the Nashville Room, Nashville Public Library, Nashville, TN).
17. James Nicholson, *Tennessee County History Series: Grundy County* (Memphis: Memphis State UP, 1982), 16; and "Wonder Cave Attracts 40,000 Visitors Yearly," *Nashville Tennessean,* April 29,1962, n.p. (photocopy on file at the Nashville Room, Nashville Public Library, Nashville, TN). The Wonder Cave Historic District was listed on the National Register of Historic Places in 1987.
18. Della Yoe, "Jewel Cave," Federal Writers' Project, 1941; Tim O'Brien, *Tennessee: Off the Beaten Path* (Chester, CT: Globe Pequot P, 1990), 16, 81. Apparently, in the late 1890s, the colonists of the adjacent Ruskin settlement opened a general store at Jewel Cave, which was then called Stalactite Cave.
19. Yoe, "Jewel Cave," 1941; Barr, *Caves of Tennessee,* 177–78. There are no documented historical ties to the Jewel Cave National Monument in South Dakota.
20. Ralph Lopard, "Ruskin Cave," Federal Writers' Project, 1941; Brandt, *Touring the Middle Tennessee Backroads,* 239–41; Robert Ewing Corlew, *A History of Dickson County, Tennessee* (Nashville: Tennessee Historical Commission and Dickson County Historical Society, 1956); and Ilene J. Cornwell, "Ruskin Colony Grounds," Nomination Form, National Register of Historic Places Inventory, 1973 (copy on file at the Tennessee Historical Commission, Nashville).
21. Lopard, "Ruskin Cave"; Clay Bailey, "Looking Backward at the Ruskin Cooperative Association," *Tennessee Historical Quarterly* 53 (1994): 100–113; and Brandt, *Touring the Middle Tennessee Backroads,* 240; Federal Writers' Project, *The WPA Guide to Tennessee,* 455–57; Ilene J. Cornwell, *Ruskin!* (N.p., n.d.), 26–28.

22. W. Fitzhugh Brundage, *A Socialist Utopia in the New South: The Ruskin Colonies in Tennessee and Georgia, 1894–1901* (Urbana, IL: U of Illinois P, 1996); Brandt, *Touring the Middle Tennessee Backroads,* 240; Clay Bailey, "Ruskin Cooperative Association," *The Tennessee Encyclopedia of History and Culture,* ed. Carroll Van West (Nashville: Tennessee Historical Society, 1998), 817–18; Francelia Butler, "The Ruskin Commonwealth: A Unique Experiment in Marxian Socialism," *Tennessee Historical Quarterly* 23, no. 4 (1964): 333-42.
23. Ann Betts, "Ruskin Cave Was Also Memorable as a College" *Tennessean,* July 10, 1985; Lopard, "Ruskin Cave," 1941; Cornwell, *Ruskin!,* 29; David Jellema, "Jazz in Tennessee," *Tennessee Encyclopedia,* 478; Charles K. Wolfe, "Music," *Tennessee Encyclopedia,* 661. The Ruskin Cave College had close ties to Trevecca Nazarene University, a training school for Christian workers, which had been established in 1901 at Nashville. In 1917 Trevecca's campus suffered from a disastrous fire, and its students and faculty apparently transferred temporarily to Ruskin Cave College. Francis Craig was related to Edwin Craig, who founded radio station WSM in 1925; his band performed at Nashville hotels and nationwide between the 1920s and 1950s.
24. Cornwell, *Ruskin!,* 26. Ruskin Cave became nationally recognized for its square dances in the 1950s and 1960s.
25. Cornwell, *Ruskin!,* 30; Lisa Human, "David Allan Coe Slates Museum for Ruskin Cave," *Tennessean,* June 18, 1982; Coe purchased Ruskin from Joe Gibson in 1982. In the late 1980s, country music entertainer Tammy Wynette considered purchasing Ruskin from Coe for use as a country music theme park, but the proposed plan never took off. Since Coe's departure, Ruskin was used briefly as a Christian home for troubled children and currently is being developed as an exotic animal park.
26. Della Yoe, "Yates Cave," Federal Writers' Project, 1941; Doug Plemons, "Other Historic Caves of Cross Plains" *Historical Quarterly: Eastern Robertson County,* Winter 1997, 8–10; Rita A. Read, *Across the Plains: A History of Cross Plains, Tennessee, 1778–1986* (Cross Plains, TN: Cross Plains Heritage Commission, 1986), 87–88; Barr, *Caves of Tennessee,* 400–401.
27. Yoe, "Yates Cave," 1941; interview with Ruby Covington, June 23, 1998; Read, *Across the Plains,* 88.
28. Eleanor Williams, *Homes and Happenings* (Oxford, MS: Guild Bindery P, 1990), 94–95.
29. Federal Writers' Project, *The WPA Guide to Tennessee,* 409; Williams, 95; and Della Yoe, "Dunbar Cave" Federal Writers' Project, 1941.
30. Federal Writers' Project, *The WPA Guide to Tennessee,* 408–409.
31. Williams, *Homes and Happenings,* 96.
32. Ibid., 97.
33. Ibid., 96.
34. Elizabeth Schlappi, *Roy Acuff, the Smoky Mountain Boy* (Gretna, LA: Pelican Publishing, 1978), 165–67.
35. Williams, *Homes and Happenings,* 98.

CHAPTER 12

PARK 'N' PRAY: AN EXAMINATION OF DRIVE-IN RELIGION IN FLORIDA

CARRIE SCUPHOLM

The history of roads is intertwined with the history of the automobile. This essay is about an unusual by-product of automobile culture. From the time of Richard M. Hollingshead's 1933 patent for a drive-in theater, the public has been fascinated with the phenomenon of traveling in the car and the ability to stay in the car and enjoy an event or entertainment at a particular destination. By the mid-1950s, about 150 drive-ins were operating in Florida. Of course, drive-ins could only be operated at night—as a result, there was an enormous amount of time when drive-ins were vacant and unused. That is, until Rev. Lowry Bowman thought of holding his church service at a drive-in theater. Climate and "snow bird" congregations dictated that this would be primarily a southern occurrence. The growing permanent population, along with the increasing number of tourists—who traveled to Florida via the Dixie Highway—put this state in the forefront of drive-in religion.

The drive-in church became part of the American automobile culture following World War II. The establishment of the drive-in church in the early-1950s coincided with the drive-in theater's peak of popularity. Drive-in churches were especially prevalent in the warmer climates of the South and West. In these days of drive-in pharmacies, banks, restaurants, and dry cleaners, few people remember the novelty of drive-in churches, which initially emerged in the 1950s.[1]

Developed to accommodate the post–World War II population boom, Florida's earliest drive-in services were held at existing drive-in movie theaters. Daytime reli-

FIG. 12.1. Worshiper at Venice-Nokomis Presbyterian Church, with window speaker. From *Life* magazine, April 18, 1955. Time Life Pictures/Getty Images.

gious services provided an interesting contrast to the nighttime entertainment at the site, a fact that did not pass unnoticed by the news media.[2] Most drive-in churches started by utilizing a drive-in theater on Sunday morning. The minister, choir, and organ were set on top of the projection booth, which was often located at the narrow end of a fan-shaped parking lot, a suitable distance from the screen, and the sermon was broadcast to cars through window speakers (fig. 12.1). Many worshipers were attracted by the lack of formality typically associated with Sunday morning church services. Encouraged to "come as you are," people arrived wearing everything from bathing suits to pajamas. Parking spaces were often filled with cars carrying boats, surfboards, and golf clubs. Churches favored the drive-in because it afforded everyone the opportunity to attend, including the sick, the physically handicapped, and families with small children.[3] In addition some people just stopped while driving by.

Prior to the 1950s, drive-in theaters had sometimes allowed churches to use their grounds at no cost, generally limited to Easter Sunday services or the occasional religious pageant. For example, in 1938 the San-Val Drive-in Theater in Burbank, California, offered its facilities free of charge to any church that wished to hold services.[4] Throughout the 1950s an increasing number of drive-ins made their facilities available to churches. By granting free use of their property to churches, drive-ins hoped to improve their image in the community. This was an industry-recommended strategy to counteract potential negative reactions to outdoor theaters that were opening soon.[5]

Florida was especially well suited for embracing the drive-in church. Prior to the popularization of the drive-in concept, Florida churches had long been confronted with the problem of accommodating large numbers of visiting worshipers each winter. One of the earliest solutions was attempted in St. Petersburg, where the Pasadena Community Church purchased additional acreage around the church and installed large outdoor speakers to broadcast the service to a crowd that spilled beyond the church walls.[6]

From the beginning, drive-in church services attracted national newspaper and magazine coverage. As early as 1951, the *New York Times* carried a story about summer church services held at an outdoor theater in Massachusetts.[7] In 1952 *Time* magazine reported on Rev. George A. Rustad's drive-in services in Phoenix, Arizona.[8] His services were offered twice nightly, five nights a week. and featured an uplifting movie- and slide-illustrated sermon. Before moving to Phoenix, Reverend Rustad attempted to hold drive-in services in South Dakota, but his efforts were thwarted by an unexpected snowstorm. Reverend Rustad was quoted by *Time* as saying, "We live in a new age, and the churches should keep moving with the time."[9] The December 1952 issue of *Popular Mechanics* featured St. John's Lutheran Church holding services at the Marker Street Drive-in Theater in Galena Park, Texas.[10] Leading the services was Rev. E. W. Matthias, who was inspired after reading about drive-in Easter morning sunrise services.[11] One year later, the trade publication *Boxoffice* ran an article titled "Church Services at Three Drive-ins Builds Goodwill in Jacksonville."[12] The *Motion Picture Herald* remarked that the Lakeland Drive-in's playground was used for a church kindergarten after services were conducted.[13]

The mechanics of most drive-in services were similar. Ushers distributed hymnals as cars arrived, helped plug in speakers, and conducted car-to-car collections during the service or asked worshipers to place donations in a bin as they were leaving (fig. 12.2). Some drive-in churches also distributed car-to-car wafers and grape juice for communion. At many drive-in churches worshipers rolled down windows and sang hymns together then congregated at the snack bar for coffee and doughnuts after services.[14]

FIG. 12.2. Ushers at Venice-Nokomis Presbyterian Church. From *Life* magazine, April 18, 1955. Time Life Pictures/Getty Images.

Drive-in church services eventually grew so popular that some established congregations acquired parcels of land and commissioned architects to design drive-in churches. Certainly the best known is Dr. Robert Schuller's Crystal Cathedral in Orange County, California. His first service was held at the Orange Drive-in Theater on March 27, 1955, where he preached from the roof of the snack bar. Although Schuller is widely believed to be the first to offer drive-in services, many of Florida's drive-in churches predate the establishment of the outdoor service begun by Schuller. Schuller's church has served as a model for drive-in churches around the country. Today his ministry is internationally televised and reaches more than thirty million viewers.[15]

According to *Time* magazine, in 1967 there were seventy drive-in churches still in operation across the country.[16] The earliest church services held at drive-in theaters were generally outreach ministries of established congregations. Once the idea proved successful, some congregations purchased land and hired architects to design ecclesiastical structures that incorporated the drive-in concept. Case studies examined here include four drive-in churches in Florida that continue to hold services today: the Whitfield Estates Presbyterian Church in Sarasota, the Venice-Nokomis Presbyterian Church in Venice, the Central Christian Church (Disciples of Christ) in Orlando, and the Daytona Beach Drive-in Christian Church.

WHITFIELD ESTATES PRESBYTERIAN CHURCH, SARASOTA

In Florida the phenomena of drive-in centers of worship had its beginnings as a vision of a Sarasota minister, Rev. Lowry Bowman. While serving as an army chaplain during World War II, Reverend Bowman led outdoor services aboard troopships. Inspired by his success preaching outdoors, he believed that a church did not need to be surrounded by four walls. After years of being forced to make use of public-address systems and double services to minister to seasonal congregations, he was convinced after seeing his first drive-in theater that it was a venue churches could employ.[17] People could and did worship anywhere, so why not a drive-in church? When he had the opportunity to organize a new church in the growing community of Whitfield Estates just north of Sarasota, he proposed this arrangement to the new congregation. The Home Mission Committee of his Presbytery encouraged him to try his idea, although they were skeptical of its value.[18] On January 1, 1952, Reverend Bowman formed the Whitfield Estates Presbyterian Church, holding services at the Trail Drive-in Theater, on U.S. 41, more or less as an experiment.[19] That Sunday morning the billboards advertised the film *The Devil Takes a Holiday*.[20] Reverend Bowman preached from a truck, with the wind intermittently knocking over the organ and pulpit. The service was an immediate success.

Shortly after acquiring a parcel of land, the congregation was able to move the services to their present location by installing the permanent drive-in features and a temporary building to house the pulpit, choir loft, and equipment.[21] The new church was dedicated December 12, 1953. The large attendance each Sunday—including some who had never before frequented church—astonished the members. Consequently the church determined that when it had raised enough money to construct a permanent sanctuary, it would incorporate the drive-in idea.[22] Four years after the church was organized, it had 325 members, with a wintertime congregation of more than 1,200 individuals.[23]

FIG. 12.3. Whitfield Estates Presbyterian Church, 7045 Tamiami Trail, Sarasota. Bail-Horton Associates, 1955–56. Photo by the author.

As the Whitfield Estates Presbyterian Church soon found, conducting a church service at a drive-in theater required several changes from traditional services. The ushers, who doubled as parking attendants, solved the problem of distributing hymnbooks by handing them out from the stand normally occupied by the ticket taker. The hymnals were returned to the ushers as people left. Regular collection plates were unsuitable because the strong winds blowing off the Gulf of Mexico whisked away checks and folded money. To solve that problem, quart ice-cream containers were substituted. Six deacons collected the car-to-car offering. On communion Sunday, Reverend Bowman asked the congregation to leave their cars and line up to receive the sacrament of the Lord's Supper.[24]

The Whitfield Estates Presbyterian Church attracted widespread attention. The *Sarasota News* carried a story in the March 12, 1955, edition, stating that photographers from both *Look* and *Time* had announced their intentions of photographing the church in session.[25] As a result of favorable publicity, other congregations in cities throughout Florida copied the idea in an attempt to attract the state's winter mobile population. It was reported that Reverend Bowman had received inquiries from all over Florida, Texas, and California.[26] Other newspaper stories talked of the United Church of Fort Myers constructing a drive-in church and the Christian Church of Daytona Beach patterning a drive-in church after Reverend Bowman's.[27] The city of Venice, just south of Sarasota, organized a drive-in church with the help of Reverend Bowman.[28]

The present Whitfield Estates Presbyterian Church was the first in the United States, so far as can be determined, to have originally been planned as both a drive-in church and a conventional place of worship (fig. 12.3).[29] It was reported that "the architectural features of this new building will be as revolutionary as the drive-in church idea."[30] Bail-Horton Associates designed the building so that the choir and pulpit are in full view of the entire congregation, both inside and outside of the building. Unlike most churches the inside congregation is seated on the second floor. A projecting steel-and-concrete canopy provides shade to the pulpit area, preventing glare. Sunday school rooms, restrooms, an office, a kitchen and a large social hall are located on the first floor. In addition to seating five hundred worshipers inside the church, there are outside drive-in facilities for four hundred cars.[31]

VENICE-NOKOMIS PRESBYTERIAN CHURCH, VENICE

Tired of meeting in a funeral parlor, the Presbyterian congregation in Venice wanted to construct a sanctuary but lacked the required finances.[32] In addition they faced a problem common to many churches in Florida's vacation destinations—large congregations during tourist season but small ones the rest of the year. After observing the widespread success of services held at drive-in theaters, the congregation decided to construct a drive-in church. The Home Mission Committee of the Presbytery purchased an 8.5-acre tract for their permanent home. Reverend Bowman, pastor of the Whitfield Estates Presbyterian Church, officiated at the new mission's organizational service on March 22, 1953, and presented the new church's charter.[33]

Architect Victor Lundy designed a temporary two-story building, measuring fourteen by twenty-nine feet. The second floor housed the pulpit, organ, and choir while Sunday school rooms occupied the first floor. Lundy sheathed the entire second story of the Outdoor Worship Center in glass so the pastor and choir would be visible from all four sides. The church grounds were attractively landscaped, and parking spaces with individual speakers were provided for one hundred cars. The help of a contractor who belonged to the congregation kept the cost to seventy-six hundred dollars.[34] The Venice-Nokomis Presbyterian Church was the second church built in the United States expressly for the worship of drive-in congregations.[35] Like the Whitfield Estates Presbyterian Church, the congregation planned for the eventual construction of indoor sanctuaries while outdoor facilities were to be retained.[36]

The congregation held their first service in the newly constructed Outdoor Worship Center on April 4, 1954, only a year after the church's organizational service and dedication (fig. 12.4).[37] It was just like a drive-in theater, speakers hung on car windows and cars lined up row on row, with families inside. The only difference was the lack of films, popcorn, and soda.[38] The church's initial lack of group fellowship and the reluctance of many people to worship in such an untraditional setting concerned Rev. Robert White. However, he soon discovered that the drive-in service

FIG. 12.4. Venice-Nokomis Presbyterian Church, Outdoor Worship Center. Victor Lundy, 1954. Demolished. Photo courtesy of Venice Archives and Area Historical Collection.

had its own unique value for "those unable to leave their cars, for the hard of hearing, and for those who avoided traditional churches."[39] In the early days attendance at the services ranged from 75 people in the summer to 275 during tourist season. *Life* magazine devoted three pages in its April 18, 1955, issue to the outdoor church and the work of Reverend White.[40] More than 325 people were present the day *Life* magazine covered the service.

In 1965, after the completion of a permanent sanctuary, also designed by Lundy, the church removed the Outdoor Worship Center from its property; it was later demolished.[41] Drive-in services have remained a Sunday morning tradition at Venice-Nokomis Presbyterian, although they are now broadcast on the radio rather than through upright speakers. Sunday morning services today include outdoor communion and a garden usher.[42]

CENTRAL CHRISTIAN CHURCH, ORLANDO

The Central Florida Drive-in Worship Service claims to be the oldest continuing drive-in worship service in the nation.[43] An outreach ministry of Central Christian Church, the first services were held on November 23, 1952, at the Winter Park Drive-in Theater. The senior pastor of Central Christian Church at that time, Rev. Paul Carpenter, saw the service as an outreach evangelical ministry, an opportunity to "reach people at the place they feel most comfortable."[44] Those who found the outdoor services convenient included the disabled, winter residents, and tourists, as well as local residents who were simply looking for a new congregational home. Some worshipers brought their dogs, and for a time, ushers handed out dog biscuits through car windows along with church bulletins. There are many stories associated with the drive-in church that have become part of the local lore. One recalls that the first time a minister got up on a concession stand to preach at the drive-in theater, someone who did not know what was happening telephoned the police.[45]

Drive-in services continued at the Winter Park Drive-in for the next twenty-eight years.[46] When the drive-in closed in 1980, the minister searched in vain to find another outdoor theater. Unable to secure a suitable location, the outdoor services were moved to the parking lot of Central Christian Church (fig. 12.5). Here, the tradition of informality continues. Rev. Stephen Gentle preaches from a mobile wooden podium at the front of the parking lot with his voice broadcast over an AM frequency. Between twenty-five and fifty cars attend the 8:15 A.M. drive-in service. Generally speaking, only a few of the worshipers are members of Central Christian Church, which still holds its traditional service later in the morning. At forty-five minutes in length, the nondenominational services and sermons are among the earliest and shortest in the area. Since the outdoor services began, nearly a half-million worshipers in more than two hundred thousand cars have attended. The invitation "come as you are, worship in your car" still stands today.[47]

FIG. 12.5. Central Christian Church outdoor worship services, 250 Ivanhoe Boulevard, Orlando. Photo by the author.

DAYTONA BEACH DRIVE-IN CHURCH, DAYTONA BEACH

On May 24, 1953, the First Christian Church of Daytona Beach established what was then known as the South Peninsula Drive-in Church Service. Rev. Glenn Murdock, minister of the church, conducted the service from the roof of the projection booth of the Neptune Drive-in Theater. That first Sunday 206 people attended in eighty-seven cars.[48]

In November 1957 Rev. Wallace Pomplun became the first regular minister of the organized church. During the early years of his ministry, the drive-in continued to operate as a movie theater through a lease with the Dixie Corporation. As a result Reverend Pomplun and his family had to arrive at the drive-in before daybreak on Sunday to clean up the bottles, bags, and garbage discarded by Saturday night patrons. The success of the drive-in service led to the church founders selling a recently purchased parcel of land in order to raise capital for a down payment on the Neptune Drive-in Theater. By January 1960 the church purchased the Neptune Drive-in for one hundred thousand dollars. During the first two years that the church owned the property, it continued to show movies to fulfill the contract with the Dixie Corporation. This required the minister and his wife to make weekly trips to Jacksonville to pick up films for the remainder of the lease.[49]

In February 1958 a "temporary" building erected against the screen became the altar and the Sunday school facility; it was used for the next nineteen years. In 1977 the old altar building was demolished, and construction of the present structure

FIG. 12.6. Daytona Beach Drive-in Church, 3140 Atlantic Avenue, altar building, 1977. Photo by the author.

was started (fig. 12.6). The altar building houses the control room, offices, a workroom, the choir room, the choir loft, and the porch that serves as the pulpit. The church enclosed the original marquee for the Neptune Drive-in in the early 1980s, and constructed a small chapel around it. The drive-in's original snack bar and restroom facility, located in the center of the eight-acre parcel, was renamed Friendship Hall. This is where the fellowship meets for coffee and doughnuts both before and after services. The snack bar menu is still visible on the exterior of the building, although layers of white paint cover it.[50]

At 8:30 and 10:00 A.M. every Sunday, services are broadcast to the congregation in Friendship Hall, to those in the Little Chapel, to cars parked in the lot, and to guests seated on benches in the shade. Today attendance averages about five hundred people with the highest attendance on Easter Sunday, when often more than thirteen hundred attend the sunrise and regular services.[51]

CONCLUSIONS

In Florida the drive-in idea was adopted as a result of the challenges presented by seasonal congregations, limited finances, and the simple fact that people like to listen to sermons from the comfort of their cars, especially those with children or invalids in the family. Furthermore, vacationers often feel they cannot go into a formal church because they are not properly dressed or they feel out of place in a strange town. Drive-in worship continues to attract many people, including the physically handicapped, the hard of hearing, tourists, and others who enjoy and benefit from outdoor services. Today the majority of people who attend are elderly. This arrangement, which allows churchgoers to remain in their cars, eliminates many of the timeworn excuses for not attending church. While the notion of drive-in churches may seem unusual, they succeed in making religion available to those whom otherwise cannot or will not attend conventional church services. There are no binding rules, except to praise God.

There are several Protestant churches in Florida that broadcast services to a designated parking area, and people can come and listen to the sermon in their cars. However, they do not retain the same old-time drive-in church concept that exemplifies the Whitfield Estates Presbyterian Church and the Daytona Beach Drive-in Church. These churches, parishioners say, represent a time when drive-in theaters were all the rage, and a time when men of vision drew on that attraction to bring religion to the passersby, the tourists, and the physically handicapped.[52]

Although Florida's drive-in churches were first planned as temporary measures to accommodate growing congregations during the post–World War II population boom, they have endured. The unanticipated benefits of accommodating the infirm, the physically disabled, and nontraditional participants have ensured that this unique tradition will continue for years to come.[53]

NOTES

1. Mary Ellen Hunter, "Drive-in Churches: The Sunbelt Phenomenon Had Its Beginnings as a Vision of a Local Minister," *Bradenton Herald,* August 1, 1998.
2. Lorrie Muldowney, "Churches Satisfy Need for Convenience," *Sarasota Herald Tribune,* February 24, 1999.
3. "Drive-in Devotion," *Time,* November 3, 1967, 83.
4. Kerry Seagrave, *Drive-in Theaters: A History from Their Inception in 1933* (Jefferson, NC: McFarland, 1992), 25.
5. Seagrave, *Drive-in Theaters,* 83.
6. "North Side Church Ends Drive," *Sarasota News,* March 12, 1955.
7. "Drive-in Theater Serves as Church on Sunday," *New York Times,* August 13, 1951, 12.
8. "Drive-in Chapels," *Time,* July 14, 1952, 60.
9. Ibid.
10. "Church Services Held in Outdoor Theater," *Popular Mechanics,* December 1952, 124.
11. Ibid.
12. "Church Service at Three Drive-ins Builds Goodwill in Jacksonville," *Boxoffice,* July 18, 1953, 23.
13. "Lakeland Plan and Policy," *Motion Picture Herald,* July 7, 1956, 15–18, 36.
14. "Drive-in Devotion," *Time,* November 3, 1967, 84.
15. Don and Susan Sanders, *The American Drive-in Movie Theater* (Osceola, WI: Motorbooks International, 1997), 108.
16. "Drive-in Devotion," 83–84.
17. "North Side Church Ends Drive."
18. Ibid.
19. "Reverend Bowman Celebrates 30 Years in Sarasota," *Sarasota Herald Tribune,* September 21, 1958.
20. Hunter, "Drive-in Churches."
21. "North Side Church Ends Drive."
22. "Sarasota Has Florida's Most Unusual Church," November 5, 1953, Whitfield Estates Presbyterian Church subject file, Sarasota County Department of Historical Resources.
23. "Whitfield Estates Church Dedication Set," *Sarasota News,* March 7, 1956.
24. Walter Powers, "Built from the Ground Up as 'Auto-Church'—Many Who Never Attended Church Attracted by Drive-in Sanctuary," *Tampa Sunday Tribune,* December 27, 1953.
25. "North Side Church Ends Drive."
26. "Reverend Bowman Celebrates 30 Years in Sarasota."
27. Ibid.
28. Ibid.
29. "Outdoor Church Drive Set," *Sarasota News,* February 19, 1955.
30. Ibid.
31. Ibid; "Drive-in Church Acquires More Land," *Sarasota News,* October 7, 1956.
32. The Venice-Nokomis Presbyterian Church was renamed the Venice Presbyterian Church on May 21, 1986.

33. Venice Archives and Area Historical Collection, Venice-Nokomis Presbyterian Church subject file.
34. Ibid.
35. "Venice Church Story to Be in Life Magazine," *Sarasota Herald Tribune,* March 15, 1955.
36. Ibid.
37. Venice Archives and Area Historical Collection, Venice-Nokomis Presbyterian Church subject file.
38. "Drive-in Church a Novelty," *Venice Gondolier,* March 20, 1993.
39. "Drive-in Church," *Life,* April 18, 1955, 175–78.
40. Ibid.
41. Venice Archives and Area Historical Collection, Venice-Nokomis Presbyterian Church subject file.
42. Muldowney, "Churches Satisfy Need for Convenience."
43. Central Christian Church, Orlando, Florida, Archives, and John C Updegraff, *The Christian Church (Disciples of Christ) in Florida* (N.p., 1981) 329–30.
44. Mark I. Pinsky, "Worshipers on Wheels: Drive-in Service Fills Need," *Orlando Sentinel,* November 22, 1997.
45. Russ White, "Worshippers Choose Car Seats over Church Pews," *Orlando Sentinel,* November 21, 1987.
46. Ibid.
47. Pinsky, "Worshipers on Wheels."
48. Daytona Beach Drive-in Christian Church Archives.
49. Ibid.
50. Ibid.
51. Ibid.
52. Hunter, "Drive-in Churches."
53. Muldowney, "Churches Satisfy Need for Convenience."

CHAPTER 13

GET RIGHT WITH GOD: HARRISON MAYES'S ROADSIDE ADVERTISING CAMPAIGN FOR THE LORD

MARTHA CARVER

Religious evangelism has taken many forms over the centuries, and it should be no surprise that with the rise of automobiles and automobile tourism evangelicals found a new medium through which to espouse their religious convictions to ever-growing numbers of people. Over approximately fifty years Harrison Mayes erected roadside signs and attractions in an effort to alert sinners to the inexorable passage of time, their immortality, and their own Judgment Day. Mayes focused on placing these signs—hearts, messages on barns or hillsides, and the ubiquitous crosses—on well-traveled U.S. routes and state highways including the Dixie Highway and other named early interstate routes. Mayes selected these sites for their visibility to the highest number of people possible and placed his messages near heavily visited tourism sites such as the Grand Canyon. Mayes would eventually place, or have others place, his signs in all fifty states, primarily in his spare time using his own money. While others have used roadside signs to convey their religious beliefs, Harrison Mayes was unique in his structured, long-term, and organized approach, which was tied to main roads rather than just his yard or local roads, as was the case with many "roadside preachers." In that, Mayes more resembled commercial advertising approaches already familiar to the motoring public, such as Burma Shave, See Rock City Barns, or Wall Drug Store, than he did religious evangelicals.

For many Christians evangelism—or taking the gospel of Christ to others—is a fundamental component of their religion and essential to salvation. This has led to a widely held evangelical doctrine that "no one ever gets to heaven alone." While few people readily associate the phrases "religious proselytizing" and "roadside architecture," roadside religious advertising is relatively common. Like other advertising, religious promotion must use a combination of bold graphics and a straightforward message to get its point across. As the historian Betty MacDowell explains, this type of roadside art and religious teaching was "intended for the perceptive eye and the receptive heart . . . to arrest the attention and speak to the spiritual needs of those who may catch only a brief glimpse as they speed by."[1]

Graphics were typically direct, such as an open Bible. However, the most common designs used a cross. Roadside crosses could take many forms, such as a single cross or a stark cluster of three crosses. These could be presented with or without a written message. If they contained written messages, the text was generally biblical verse or references to a specific verse. Signs, sometimes in a series reminiscent of the Burma Shave signs or painted on sheds or barns, were short, simple, pithy, and to the point. Frequently people erected informal signs on their own property, but occasionally a folk artist or minister erected his artwork on the property of others. The people who erected these signs tended to possess an unshakable faith, a boundless will to work, and a vision that no one else saw.

One of the most prolific itinerant artisans of religious roadside advertising was Henry Harrison Mayes. Born in 1898 in the Appalachian coal country of rural East Tennessee, Mayes lived in Claiborne County until the 1940s. His father originally farmed, but when Harrison was about ten, the family moved to the town of Fork Ridge, Tennessee, where his father became a coal miner. Young Mayes completed the fifth grade, and by the time he was thirteen, he had begun working in the mines for the Fork Ridge Coal Company. Mayes worked as a coal miner for the next forty-three years. In 1946 he moved to nearby Middlesboro, Kentucky, where he lived until his death on March 7, 1986.[2]

When Mayes was nineteen, he married Lillie Minton.[3] Originally he was associated with the conservative Church of God, but at times he could be fiercely independent in dogma and later formally renounced his church membership to pursue a more individualized quest for salvation. Recognizing an unwavering conviction for serving God, Mayes considered various options at an early age. After reading in a church magazine that millions of people never go to church or read a Bible, Mayes pondered how to convey the gospel of Christ in a nontraditional format.[4] Roadside advertising was his answer. At age eighty-six, he recalled, "I tried preaching, I tried running revival meetings, and I tried going all over Florida and different places making music. I never could make a go at nothing until I got into this sign work. That was exactly my calling."[5]

FIG. 13.1. Mayes spent a lifetime erecting roadside crosses, but he returned to his first love, signs, for the last message he erected before his death. Mayes and his youngest son, Clyde, built the large banner-type sign in 1978 by joining four 4' x 10' panels to a wooden frame. The sign is located off State Route 33, about one mile south of New Hope Baptist Church, south of Tazewell, Tennessee. Photo by the author.

Mayes began his roadside ministry in the 1910s. Reputedly his first sign, which said SIN NOT, was written on both sides of the family's free-ranging pig.[6] Beginning when he was about fifteen, Mayes wrote messages on logs that he put "all over the place."[7] He also nailed handwritten cardboard signs on telephone poles. He soon hired a local printer to make more-formal placards. He also wrote messages such as JESUS IS COMING SOON, PREPARE TO MEET GOD, or JESUS SAVES on oilcloth banners, in red paint on rocky hillsides, and with chalk on railroad cars.

Mayes quickly realized that these media were too ephemeral and inefficient. To reach more souls, he recognized the need to place his work where large numbers of people would see it. As businessmen before and after him, he came to see the roadside as a perfect pulpit for a captive audience that was looking for interesting diversions. His first true roadside sign, and the first type that he mass produced, was a reversible wooden slat sign formed from three horizontal slats framed within a rectangle that read JESUS IS SOON COMING and PREPARE TO MEET GOD.[8]

A life-threatening accident in the coal mine in the late 1920s changed Mayes's life. With doctors not expecting him to live, Mayes prayed for God's grace and healing—promising to dedicate his life fully to God and his ministry should he live. He placed on himself the mission to "properly advertise the Second Advent and the wages of sin on land, on sea, in the air, and on the major planets."[9] From this experience Mayes came to the conclusion that the best method for him to serve God was to place signs throughout the United States that admonished people to repent of their sins and to accept God, a goal from which he never wavered.[10]

FIG. 13.2. This barn is located on U.S. 70 (Dixie Highway) east of Newport, Tennessee. It contains the message PREPARE TO MEET GOD in duplicate attached to the barn. Mayes placed the bottom sign, formed by corrugated metal, in the 1950s and later returned to place the second sign, which is made of wood. Photo by the author.

As a poor coal miner in the 1940s, Mayes supported his wife and children on an income of two hundred dollars a month. It took money and time not only to buy materials and make the signs, but also to pay the cost of traveling throughout the country to erect them on weekends and vacations. By the time he was eighty-five, Mayes estimated that he and his wife had spent at least seventy-five thousand dollars of their own money on their evangelical calling. However, he also credited friends and businesses with contributions of time, money, and materials to his cause.

One of Mayes's most creative funding mechanisms was through the assistance of coal mine officials. The company paid miners by the load and assigned each miner a number against which loads were credited. Mayes asked the company for an extra number that he called his "Sermon Number." When Mayes worked overtime or on holidays, he credited the loads to his Sermon Number. Other miners could contribute to his evangelistic work by crediting their loads to the Sermon Number whenever they wished as well.[11]

Mayes seemed to have been alert to innovative advertising venues, and typical of many folk artists, he frequently used recycled materials. In 1934 Mayes hung a huge sign in a gorge along the highway south of Harrogate, Tennessee. Using metal roofing from the Fork Ridge Coal Company powerhouse that the company was tearing down, Mayes cut out four-foot letters for the reversible sign that read PREPARE TO MEET GOD and JESUS WILL SAVE YOU.[12] The same year Mayes used an old boiler discarded by the coal company for another U.S. 25E sign nearby. Mayes placed the round boiler next to the roadside and painted messages on its five-foot ends. One end contained a white heart emblazoned with black letters saying GIVE GOD YOUR HEART.[13]

Similarly, Mayes retrieved and recycled an armature discarded by the machine shop. He painted the rectangular metal object white with red hearts atop the large bolts—certainly an attention getter. Around the hole in the center of the armature, Mayes wrote *Sacred Atomic Inergy* [*sic*] *Inside.* On the other side, he wrote the irresistible *Please Don't Look in This Hole* (with big arrows pointing at the hole). Inside

FIG. 13.3. This cross near Rockwood, Tennessee, on U.S. 27 (Dixie Highway) contains the messages NOTHING BUT THE BLOOD and JESUS CHRIST IS LORD. Although this cross closely resembles the work of Harrison Mayes (even though the messages are different from Mayes's typical signs), Rockwood resident Donnie Bohanan erected this sign. Photo by the author.

the hole were the words *Trust in the Lord.* Since it was inappropriate for typical roadside display, Mayes hit on the next best thing by placing it in the Greyhound Bus Depot in Middlesboro, where it was prominently exhibited in the barbershop.[14] Another of his unique roadside advertising motifs was a concrete bench along Cumberland Avenue in downtown Middlesboro. The six-foot-long bench featured a recessed panel with the words GET RIGHT WITH GOD.[15] Over the years, Mayes also painted his messages on barns. An example from the 1950s that includes the message PREPARE TO MEET GOD in four-foot metal and wood letters on the gable end still exists on U.S. 70 near Newport, Tennessee (fig. 13.2).[16]

Mayes lived most of his adult life in Middlesboro, one of the few cities in the world built in a meteor crater, a deep circular depression about 3.7 miles in diameter with ridging slopes ranging as high as 19 feet.[17] Mayes used this unique topography to create one of his best-known local landmarks, the Lighted Cross. Erected by 1949, it lies flat against a steep hillside and is visible at night from the town's main street. Formed of lines and bulbs, the cross is 120 feet tall, and the crossarm section is 60 feet wide with eighty-two light bulbs. For a time Mayes paid the electric bill; later, however, local residents donated money to help pay the bill, and friends maintained it. After a tornado severely damaged the Lighted Cross in 1988, it was rebuilt.[18]

Mayes did not limit himself to the typical roadside signs, but built "air" signs as well. About 1945 he erected a giant sign on a hillside beside his house in Fork Ridge, using three hundred tons of large creek rocks painted with whitewash. Probably inspired by World War II victory gardens, the sign read V—IN GOD. The family planted its garden in and around the letters, which could be seen only from the air.[19] Mayes also erected large metal signs near airport runways, positioned so passengers would see them as their plane took off. The signs typically read PREPARE TO MEET GOD. Airport authorities believed they upset the passengers, however, and usually disposed of them quickly. It is known that Mayes erected twenty-five to thirty such signs at airports in Little Rock, Memphis, Knoxville, St. Louis, Chicago, Cleveland, Philadelphia, Atlanta, Miami, Raleigh-Durham, and West Palm Beach. The Knoxville airport removed at least three of the signs, and the Tennessee State Museum has preserved one airport sign in its collection.[20]

Mayes brought roadside advertising into the construction of his home. He and his wife, Lillie, moved in 1946 from Fork Ridge to Middlesboro, where they built a four-room house in the shape of a cross, which they called the "Air Castle."[21] An ongoing project, Mayes expanded the house over the years, and it eventually came to be called "The House of Many Crosses." The yard and the house, which Mayes built of handmade blocks that each contained a small cross-shaped indentation, featured many elements with biblical symbolism. Twelve windows in the front section

FIG. 13.4. This unusual cross is located along the roadside at Mayes's home in Middlesboro, Kentucky. It reads LOST FOREVER BECAUSE I FORSAKENED [sic] THE CROSS OF JESUS CHRIST SAID REGRET. The reverse reads THANKS TO GOD FOREVER FOR LANDING ME SAFE THROUGH THE CROSS OF JESUS CHRIST. Photo by the author.

represented the twelve apostles, and ten windows on the rear section represented the Ten Commandments. Eight outside doors represented the eight people saved in Noah's Ark. The metal roof originally contained the message GET RIGHT WITH GOD, although Mayes would later add JESUS SAVES. A concrete header above the windows in front read JESUS SAVES.

The house contained three chimneys, representing Jesus and the two thieves at the crucifixion. Concrete fence posts represented the seven continents. Eight metal posts and one wood post, called the Judas post, represented the nine planets. The lot had twelve corners, representing biblical patriarchs and the apostles. Mayes poured a row of twenty-eight concrete hearts and crosses in the backyard, which would be moved to different locations in later years. Mayes's house is substantially altered today. Two large crosses flank the roadside; they resemble his standard roadside crosses but read THANKS TO GOD FOREVER FOR LANDING ME SAFE THROUGH THE CROSS OF JESUS CHRIST and LOST FOREVER BECAUSE I FORSAKENED [*sic*] THE CROSS OF JESUS CHRIST SAID REGRET (fig. 13.4).[22]

Although Mayes erected a wide variety of signs, he would become most famous for his concrete crosses, which he made himself in a workshop at home. He said that by 1949 he had erected more than two thousand signs (presumably cardboard, wood, or concrete) in twenty-five states and that friends or ministers (to whom he shipped signs) had erected them in all forty-eight states.[23] In an interview at age eighty-six, Mayes said that he had personally erected signs in forty-four states and had provided for the distribution of messages in all fifty states and eighty-two nations.[24] Examples of his work are in the collections of the Smithsonian Institution, the Museum of Appalachia in Norris, Tennessee, and the Tennessee State Museum in Nashville.

Mayes originally erected wood crosses, but he began experimenting with concrete in the early 1940s. These crosses cost about twelve dollars each to make. Substantial in size, each was approximately eight-feet tall and weighed about fifteen hundred pounds.[25] Originally making them four inches thick, Mayes later expanded the depth to six inches, and later still to eight inches. At Fork Ridge the sand and gravel that Mayes scavenged from the railroad tracks contained coal particles; after moving to Middlesboro in 1946, he was able to use sand and gravel free of coal. Most examples are in the shape of a cross, but some signs contain hearts on a post. Each side contained a different inscription, depressed within the cross and often painted red to symbolize the blood of Jesus. The most common messages that Mayes used on his crosses were PREPARE TO MEET GOD, JESUS IS COMING SOON, and GET RIGHT WITH GOD (Fig. 13.5).

Overtime and increased wages at the mines to accommodate supply production increases for World War II enabled Mayes to expand his production of concrete

FIG. 13.5. A heart-shaped JESUS IS COMING SOON sign in Harlan, Kentucky, on old U.S. 421. Photo by the author.

crosses substantially, many dating from the 1940s or later.[26] His work in Smyrna, Tennessee, is typical; at an unknown date Mayes had erected a wooden cross on U.S. 41/Dixie Highway. As he often did, Mayes also provided a concrete cross to a local congregation, here the Smyrna Church of God. In the early 1940s the congregation replaced the deteriorated wood cross with the concrete version (fig. 13.6).[27] In 1949 Mayes told a reporter he had erected eighty concrete signs along the highways in six mountain states and planned to replace some of the wooden crosses in other states as soon as his finances allowed.[28]

As a young man, Mayes owned a car and later an old ambulance that he used to haul his concrete crosses, though neither worked well. Mayes also hired a truck and driver, or asked for volunteers, to deliver the signs. During the early years he slept in the truck in a big wood box that doubled as a tool chest, or under the truck; but in later years the drivers and Mayes stayed at motels.[29] He usually made two sign trips a year, typically lasting ten days to two weeks. In 1963, for example, Mayes left in a hired truck with seventy-five GET RIGHT WITH GOD signs that he planned to erect at Yellowstone National Park and the Grand Canyon.[30] At the height of the automobile tourism era, Mayes must have hoped to reach an extensive audience by capitalizing on the high volume of visitors at these tourist meccas.

FIG. 13.6. This early 1940s cross in Smyrna, Tennessee, replaced an earlier wooden cross. In recent years the city improved this intersection, and the sign was relocated off the immediate right-of-way and refurbished. Photo by the author.

Mayes returned to his first love, roadside signs, for his last installation. He and his youngest son, Clyde, erected a large banner south of Tazewell, Tennessee, in 1978. Consisting of four ten-foot panels, each four-feet high and nailed to a wooden frame, the sign reads GET RIGHT WITH GOD, lettered in black on a white background. The sign remains just off State Route 33, about one mile south of New Hope Baptist Church.[31]

Like many others, Mayes believed his commitment to God was greater than the laws of humanity, and he periodically met opposition from authorities. In a 1949 interview he stated that his two greatest opponents were the American Automobile Association (AAA) and the Commonwealth of Virginia. AAA officials had testified at a senate hearing of "cranks who take a ghoulish glee in spreading" a message of violence and disaster and cited an example of a man who suffered a car crash trying to read a sign on an underpass that read PREPARE TO MEET GOD—a Mayes sign.[32] Virginia took a hands-on approach and removed thirty of Mayes's early wood signs from its roadsides, then sent him a bill for the labor. When Mayes refused to pay, Virginia threatened court action. A compromise, or perhaps more accurately, a stand-off, ensued in which Mayes did not pay the bill and lost his signs, but he stopped erecting them in Virginia.

In a 1970s interview Mayes recounted that he had recently received a letter from the Ohio State Police informing him that since his signs were on state right-of-way, they would have to be removed. He asked Ohio to leave the signs as long as possible or until he could get there—never intending to go. Just two days after returning from a trip where he installed a sign near Yellowstone, state officials wrote him that it had been removed in compliance with Lady Bird Johnson's beautification law.

In an interview during the 1980s, Mayes recalled that over the past forty years owners had rarely given him permission to place signs on their private property; and when they denied him access, he erected the signs when they were not looking.[33] Mayes placed the signs in what he called "no man's land," fencerows where ownership was cloudy and where the state might think the sign was on the farmer's land, or vice versa. He trespassed—knowing he was trespassing—because he felt it was his "religious right" to put the signs "out of the way" where they could be seen by travelers but would not cause harm to motorists.[34]

Today this uneasy relationship continues. Some states have outlawed roadside shrines although other states view them as a silent reminder to drive carefully or not to drink and drive, and thus they serve a positive transportation function. The states that oppose such signs cite safety as an issue since the markers themselves can pose a life-threatening hazard to motorists. Since many such signs rely on cross-related graphics, religious plurality has also led to opposition. As road widenings occur around the country and the passage of time has resulted in deterioration, the remaining signs created by Henry Harrison Mayes have become increasingly threatened. Family members are attempting to identify his signs and relocate the crosses off state rights-of-way and onto private property visible from the roadside (fig. 13.7). The author has identified extant signs in Georgia, Tennessee, Alabama, and Kentucky, but since Mayes erected them around the country, some may still exist in other locations.

Like "See Rock City" barns and Burma Shave signs, religious advertisements such as Mayes's crosses are a part of the roadside experience. Spotting one on the road brings to mind the man who spent his vacations erecting them.

FIG. 13.7. This early 1940s concrete cross was originally located south of Harrogate, Tennessee, on U.S. 25E (Dixie Highway). In 1999 the Mayes family relocated it to U.S. 25E in Harrogate at the Burchett Car Lot across from Lincoln Memorial University. Photo by the author.

NOTES

This essay was reprinted from *SCA Journal* 18 (Fall 2000): 14–19. Used with permission.

1. Betty MacDowell, "Religion on the Road: Highway Evangelism and Worship Environments for the Traveler in America," *Journal of American Culture* 5 (Winter 1982): 63.
2. Eliot Wigginton and Margie Bennett, eds., "Watchman on the Wall," *Foxfire 9* (NY: Doubleday, 1986), 329; Catherine Mayes, *A Coal Miner's Simple Message* (Middlesboro, TN: Clyde and Catherine Mayes, 1999), 24.
3. Wigginton and Bennett, "Watchman on the Wall," 333; Mayes, *A Coal Miner's Simple Message,* 121.
4. Dean Cadle, "The Man behind the Signs," *Louisville Courier-Journal,* January 23, 1949.
5. Wigginton and Bennett, "Watchman on the Wall," 333.
6. Fred Brown, "An American Way of the Cross," *Knoxville News Sentinel,* April 11, 1998; Scripps Howard News Service, www.reporternews.com/religion/cross0411.html (accessed January 3, 2000).

7. Wigginton and Bennett, "Watchman on the Wall," 333; Byron Crawford, "God's Ad Man . . . Kentuckian Erected A Multitude of 'Get Right with God' Signs," *Louisville Courier-Journal,* November 16, 1983.
8. Crawford, "God's Ad Man"; Mayes, 26–27, 115; Wigginton and Bennett, 333, Bill Woolsey, "Highway Evangel," *Nashville Tennessean Magazine,* July 17, 1949, 5, 8, 10.
9. Woolsey, "Highway Evangel."
10. Mayes, *A Coal Miner's Simple Message,* 5–6; Wigginton and Bennett, 329.
11. Cadle, "The Man behind the Signs"; Crawford, "God's Ad Man"; Mayes, *A Coal Miner's Simple Message,*15, 30.
12. Ibid., 24, 83.
13. Mayes, *A Coal Miner's Simple Message,* 23, 105. It is not known what was written on the other end.
14. Ibid., 37–38.
15. Ibid., 39.
16. Marcel Klar, telephone interview with Martha Carver, July 7, 2000.
17. Rick Christopher Lasley, "Middlesboro May Be Only City in World Built in Meteor Crater," *Middlesboro Daily News,* September 30, 1986.
18. Mayes, *A Coal Miner's Simple Message,* 90–92; Woolsey, "Highway Evangel."
19. Mayes, *A Coal Miner's Simple Message,* 31, 33.
20. Eleanor Dickinson and Barbara Benziger, *Revival!* (NY: Harper and Row, 1974); Mayes, *A Coal Miner's Simple Message,* 75–76.
21. Woolsey, "Highway Evangel."
22. Cadle, "The Man behind the Signs"; Dickinson and Benziger, *Revival!,* 174; Mayes, *A Coal Miner's Simple Message,* 101–2; Wigginton and Bennett, "Watchman on the Wall," 331; Woolsey, "Highway Evangel."
23. Cadle, "The Man behind the Signs."
24. Wigginton and Bennett, "Watchman on the Wall," 333, 342.
25. Mayes, *A Coal Miner's Simple Message,* 30.
26. Cadle, "The Man behind the Signs."
27. Aubrey Tune, interview with Martha Carver, April 10, 1994.
28. Cadle, "The Man behind the Signs."
29. Cadle, "The Man behind the Signs"; Crawford, "God's Ad Man . . ."; Mayes, *A Coal Miner's Simple Message,* 22, 43.
30. Mayes, *A Coal Miner's Simple Message,* 55.
31. Ibid., 34.
32. Cadle, "The Man behind the Signs"; Woolsey, "Highway Evangel."
33. Dickinson and Benziger, *Revival!,* 166–70.
34. Wigginton and Bennett, "Watchman on the Wall," 337.

CHAPTER 14

BUY ONE, GET TWO FREE: FIREWORKS

KARL PULJAK

Fireworks are not unique to the South, but they are more prevalent there. This is partly because a few southern states have allowed year-round sales of fireworks. People heading south from states where fireworks were not sold or were sold only seasonally would jump at the chance to buy fireworks and take them home—whether they were legal or not. Location of the stands is paramount and, when the Dixie Highway was the main route south, that meant that every type of stand could be found along the road. Years of survey work have allowed the author to formulate an original typology of fireworks stands. In contrast to the study of hotels and motels, caves, or drive-ins along the Dixie, the fireworks stands may have been architecturally unimposing, but they managed to get more bang for their bucks.

Like many kids in my St. Louis, Missouri, neighborhood, I had a passion for two things: summer vacation and explosives. The closing of schools in early June opened the doors to three full months of baseball-playing, grass-cutting, barbecue-eating, and pool swimming abandon. Clearly the highlight of the summer was the Fourth of July. Anticipation of that day rendered the entire month of June a form of leisure purgatory for me. It appeared to me that life in the summer seemed to hang its hat on this particular day. Of course, this holiday was laden with the usual suburban rituals at my house (backyard barbecue, a friendly game of lawn darts); however, none was more thrilling than my father's taking us across the bridge of the Meramec River into Jefferson County to buy fireworks. Sure, there were huge public displays sponsored by groups such as the Lion's Club in town, but nothing was as good as attempting to make our own fireworks extravaganza in our backyard. The moment our Oldsmobile station wagon crossed the river, as we left St. Louis County and entered the town of Arnold, over in Jefferson County, my brother and I could see fireworks stands pop up from the landscape like mushrooms: a juvenile's Shangri-La. These long, wooden structures were the warehouses of our pyrotechnic dreams.

IT SOMEHOW ALL COMES FULL CIRCLE

Although my fascination with bottle rockets and roman candles has waned over the years, I have returned to these temples of my youth, for more reasons than I can possibly explain. Certainly my fascination with our contemporary, consumer culture is one. However, my current work as an architecture professor has reminded me to be aware of and responsive to all elements of the built environment, especially buildings and spaces rendered as ordinary or unimportant: this as a way to make my students aware of the subtleties of a place. A more primary reason, however, is an inherent need to connect my own personal history to a larger context. I don't think this pursuit is an ironic deconstruction of the type of suburban childhood that many of my fellow generation-xers rant against. Yet in retrospect there was a banality prevalent in the areas where many lived as children: the houses in their subdivisions, the grocery stores, strip malls, fast food restaurants, parking lots, billboards, and traffic lights rendered the landscape a static expanse. Perhaps that's why that four-mile drive to the fireworks stand, to another place, affected me so much.

WHY FIREWORKS STANDS?

Offering a framework and typology of the U.S. fireworks stand, one of the more curious roadside building types on our landscape, I would like to contend that the stands are different from any other vernacular building type on the roadways of the South. Well, maybe they're not . . . perhaps I've been looking at them for such a long time that I've promoted them to a self-perceived importance. These structures and their evolution are more culturally meaningful than they may initially appear. They are simple yet bold objects in the landscape that transcend a set formal and structural appearance. The majority of fireworks stands are transient or simply sit on the land for months at a time, unoccupied. They have evolved over time to embrace cultural and economic change and have regional and even localized tendencies.

I have waxed poetic, and I have analyzed typologies regarding these structures. After all of this posturing and speculating, the thing that I have come to understand about them is that they simply *are*. Fireworks stands are neither designed by architects nor constructed in any particular outstanding way. Many are simply enclosures occupied by fireworks, while other fireworks stands are mass-produced. Many of those have been "particularized" or "retrofitted" in some way to personalize their use more expeditiously. Some are even outstanding examples of folk architecture. At first glance there is a latent banality about them that I appreciate; yet an idiosyncratic presence persists.

Presented here are my own observations, based on a visual survey of fireworks stands, as to the evolution of this building type. Again my attempt is not to poeticize or overanalyze these structures. It is more important to realize that they are a condition of circumstance that is larger than their own making. Fireworks stands are inherently tied to the U.S. landscape. They are cheap, at times they are peculiar, and they work. There is an intrinsic American ethos about them. It is essential to realize that their construction is not only predicated on consumption, but the buildings themselves are a product of consumption.

DETERMINED BY LAWS

One of the more unique characteristics of fireworks stands that distinguishes them from other commercial enterprises is that their location, appearance, and very existence are a product of state, county, and local laws. That these buildings are only allowed to be open for business for the equivalent of about 10 percent of the calendar year is a primary factor in understanding how and why these structures appear as they do. With the notable exception of states such as Tennessee and South Carolina, where fireworks can be sold throughout the calendar year, nearly all states that permit the sales of Class C or consumer fireworks allow these businesses to remain open for approximately three weeks during the Independence Day holiday and an additional two weeks over the Christmas/New Year's Eve holidays. On conclusion of these selling seasons, fireworks stands are emptied of merchandise and then boarded up or taken down. Depending on the circumstance, the structures are either moved to a storage lot or left dormant on the landscape. For approximately 90 percent of the year, we see only traces of their brief and vibrant activity, as they lie dormant—bold, silent monuments along the roadside during their eleven-month hibernations.

Because of their relatively short annual half-life and the peculiarities in state, county, and local laws, location becomes a key factor in the financial success of fireworks stands. Geographic and political borders, as well as moderately prominent crossroads, become important siting considerations. As a means of pitting the

FIREWORKS

The exact origin of fireworks remains vague; however, most historians agree that they were discovered in China between 800 and 1000 C.E. One legend states that a cook accidentally discovered the explosive power of a mixture of sulfur and potassium nitrate inside bamboo. According to another legend, monks searching for a potion for eternal life mixed the ingredients, causing them to explode. No matter who discovered fireworks they were, early on, seen as potential weapons and potential entertainment. As a weapon, fireworks were first used as a scare tactic against invaders, with actual bombs quickly following to blow up enemies rather than to scare them. The use of fireworks spread to the Middle East and Europe through invasions and trade. Some accounts state that the Mongol invasions of China possibly brought them to the Middle East and Europe, with other accounts crediting both Marco Polo and the crusaders for bringing fireworks to Europe.

Although fireworks had been used for ceremonial and entertainment purposes since their invention, Europeans developed more powerful and prettier examples. In the eighteenth century European royalty became known for using fireworks at lavish celebrations. One of the better-known examples is the celebration at the end of the War of the Austrian Succession, when George Frideric Handel composed *The Musick for the Royal Fireworks*. Peter the Great of Russia announced the birth of his first son with a fireworks display, and the English celebrated the coronation of Anne Boleyn with a display. It was not until the nineteenth century, when color was added, that fireworks became an increasingly popular form of entertainment. The addition of metals such as sodium, copper, and barium caused fireworks to have sudden flashes of bright reds, greens, yellows, and blues. Adding magnesium, aluminum, and titanium to fireworks created their modern appearance of brilliant flashes of light and color.

Much like their European counterparts, Americans have used fireworks since the seventeenth century as both weapons and entertainment. Most often associated with the Fourth of July celebration, fireworks have been used to mark Independence Day in the United States since 1777. During the twentieth century the mass production of fireworks made them more readily available to a wider array of customers. This widespread availability led to the creation of specialty shops and roadside stands devoted solely to fireworks. As more amateurs gained access to fireworks, the incidence of accident and injury increased, leading the government to regulate the sale and use of amateur displays. Although these restrictions lessened their use by amateurs, the fireworks industry generates millions of dollars in the thirty-two states where fireworks continue to be sold to the general public.

TAMMY ALLISON

"haves" against the "have-nots," setting up a fireworks stand just across the border from a city, county, or state where the purchase of fireworks is illegal is a typical means of stirring up business. Several examples of this pattern exist along the Dixie Highway in Marion County, Tennessee, near the Alabama state line. Likewise, primary and secondary crossings of state and county roads allow an opportunity for a driver and potential customer to slow down to observe the activity on the side of the road.

These stands are open for business during both the hottest and coldest times of the year, so their placement within easy reach of an electrical supply from a utility pole becomes vital, providing the necessary lighting and climate control for summer and winter employees. The temporary structures become like parasites, living on borrowed time in a borrowed context.

ON TYPOLOGY, TECTONIC AND FORM

Clearly the legalization of year-round sales of fireworks in a few states has allowed certain structures to become attached to the ground and to possess all the comforts of other commercial outlets: centralized air-conditioning and heat, not to mention other features of "real" buildings such as doors, floors, and walls. Because these buildings are used year-round, they can house a larger quantity and greater variety of goods, including nonfireworks goods such as souvenirs and food. Many of these businesses are run by corporations with fleets of mobile, franchised structures waiting in the wings.

While many commercial roadside building types are simply variations of an indoctrinated form, fireworks stands use of several construction materials and methods, which refutes the idea of a prescribed building type. I am not suggesting that because of the variety of forms they take, they aren't recognizable as a defined type in the landscape. However, over time and based on the locations of various structures, it is clear that fireworks stands have evolved into many forms with inherently different structures. Detailed below are typologies based on architectural surveys of fireworks stands.

PLYWOOD BOX

The plywood structure seems to be one of the more prevalent types of fireworks stands, particularly in the southern and western regions of the United States (fig. 14.1). Framed by wood studs and sheathed entirely in plywood, these structures are similar in construction and form to many seasonal fruit stands that dot the roadways near orchards, gardens, and farms. The operable front flaps of this opaque box are propped open to view and purchase products, as well as providing a visorlike shading device for both consumer and seller during the over-the-counter transaction. Reliant entirely on the American, off-the-shelf modules of the two-by-four wood stud and the four-foot-by-eight-foot sheet of plywood, the structures themselves base their overall dimensions on the material from which they are constructed. They are four or six feet wide by sixteen or twenty-four feet long and eight feet tall. By relying on the modular unit, these stands can be efficiently premade, shipped to a site, and

FIG. 14.1. Plywood box fireworks stand, Texas. Photo by the author.

dropped into place. Often several units are delivered to a site and are later attached to form a longer structure.

The fireworks stands contain built-in shelving to display the wares of the seller clearly. Only the seller and the goods occupy the box. The consumer remains outside, viewing the wares through the open plywood flap. The seller serves as an intermediary between consumer and goods, making the interchange between the buyer and seller somewhat formalized in the plywood box structure: the seller is the overseer of the goods, and the buyer is the passive inquirer. Any transaction relies on the inquiry of the purchaser through the proprietor, who allows the transfer of items from the shelves of the stand into the hands of the consumer.

In addition to providing an efficient means of enclosure, the plywood walls serve as a canvas, an easy surface to paint quickly and thoroughly. Applied color, text, and images animate these rather simple structures. The application of the usual color trilogy of fireworks stands—white, red, and yellow paint—transforms the buildings into vibrant, graphic, and bold statements on the landscape, legible from moderate distances and at moderate rates of speed from passing automobiles. Overhead signage and decorative lighting serve as additional low-tech ornament, attempting to draw in passersby. At the conclusion of the selling season, the plywood boxes are either removed from their site and transferred to a holding facility or remain as empty shells along the roadside.

TENTS

Prevalent in the Midwest, tent structures are an efficient means of displaying and selling fireworks (fig. 14.2). The relative economy in erecting and striking the structure is one of the most notable characteristics of the tent, which is certainly advantageous for an enterprise that commences and concludes yearly operation in three weeks. Within a matter of hours one can enclose a larger volume of space. The ideal quality of such a nomadic structure is the tent's inherent ability to collapse and fold into a relatively small package, allowing the seller not to be burdened by the storage of a larger rigid structure at the conclusion of the selling season. To further expedite this activity, several tent companies rent their structures to fireworks sellers during the selling season. Trucks filled with canvas, stakes, and rope convoy across the country, pitching tents in predetermined locations along the way. Clearly the temporary and nomadic qualities of tent structures have been tied to the tradition of many festival cultures, such as circuses and carnivals, by providing a transient space.

FIG. 14.2. Tent fireworks stand, Arkansas. Photo by the author.

Because of this much larger area to house and sell goods, a primary advantage of these structures over the plywood stands is that a greater quantity and variety of goods can be displayed. The concept of more choices, or at least the illusion of more choices, becomes a method of luring would-be buyers. A means to arrange the display of goods requires the need for additional props such as larger folding tables, boxes, and baskets. This equipment must be organized in such a way to ensure sufficient circulation paths for shoppers. Unlike the restricted access to goods in the plywood box setup, the display stands and aisles in the tent structures allow consumers to browse and to handle merchandise in relative privacy, a shopping experience with which most people are comfortable. With this more unrestricted access to goods for the shopper, issues of entrance, exit, and edge must be clearly delineated not only to inform and orient the consumer, but also to optimize the surveillance of the marketplace by the merchant.

TRAILERS AND TRUCKS

Mobility and transience have become the primary contributing factors in another form of fireworks stands (figs. 14.3 and 14.4). Trailers remain a vital by-product of the U.S. road—small buildings on wheels. The structures are simply pulled to a location, detached, and dropped off. Recreational campers and tractor-trailers are the clearest prototypes of this means of carrying and transporting space, people and goods. I have seen many examples of pyrotechnic entrepreneurs who systematically retrofit these types into buildings for the sale of fireworks. Additions of stairs, windows, shelving, air-conditioning units, and signage transform the trailers into a system of storing and displaying goods. Goods are both carried to the site and displayed in the same container, allowing for minimal amount of setup time. Purchased as "as-builts," the trailers lose some of their ubiquity once the fireworks seller sets up shop and uses signage to lure passersby.

Similarly the trailer of an eighteen-wheeler, so prevalent on the highway, provides an even larger volume of mobile space to store and display goods. These trailers can be driven to a site, unhitched from the cab, and left on the landscape. Again the additional equipment to provide access and signage are then attached to the trailer to assist in its function as a retail venue. Mobile homes, while not as prevalent a fireworks building type, have also been used in a similar manner. Built for human habitation, these structures provide many more modern comforts that most traditional stands lack, such as plumbing and air-conditioning.

FIG. 14.3. Travel trailer fireworks stand, Texas. Photo by the author.

FIG. 14.4. Eighteen-wheeler fireworks stand, Alabama. Photo by Adam Bailey.

Like the plywood box form, these structures are typically rectilinear. However, the form of the trailer is not determined by a construction module, but by the fact that it is pulled or towed by a truck. And, unlike the plywood stands, many of the trailers are actually wide enough to allow the consumer to inhabit the space of the trailer itself during the transaction. For example, the eighteen-wheeler fireworks stand is organized so that the consumer actually enters through one end of the trailer. The smaller travel trailer relies on the plywood box's buyer/seller relationship

METAL BUILDINGS

The metal-building industry has contributed significantly to the built environment along the U.S. roadside (fig. 14.5). A cheap and expedient way to provide raw space, the industry that produces it has become an undeniable force, manufacturing the medium-to-large structures found along commercial strips and in industrial parks. The metal-building industry has also mass-produced smaller buildings for noncommercial or nonindustrial use, such as backyard utility or garden sheds and other small storage structures. These structures are also used as fireworks stands, primarily located in gas station parking lots. Lighter, easier to maintain, and less cumbersome than plywood stands, metal buildings are one of the more cost-effective structures.

Metal buildings are mass-produced in factories for general use, making post–assembly line alterations to customize this ubiquitous structure into a fireworks stand inherently more difficult than it is for other structures previously

FIG. 14.5. Mobile home fireworks stand, Alabama. Photo by Adam Bailey.

described. For example, painting bold graphics onto the metal surfaces of these buildings does not replicate the quality of the original factory paint job. As a result, signage is typically not directly applied to the metal building but is attached by bolts or ties to it. In a locale where the sale of consumer fireworks is only allowed for a brief time, these versatile metal structures can be pulled out and quickly filled with merchandise, and signage is easily attached. Likewise, at the conclusion of the selling season, the metal shed can be expediently emptied and readied for another use for the remaining eleven months of the year.

PERMANENT BUILDINGS

With the year-round sale of consumer fireworks in some locations, large permanent buildings have been constructed on purchased land to facilitate this need (fig. 14.6). Often dubbed with larger-than-life self titles such as "wholesale warehouse outlet" or "world's largest" as a part of their company names, these structures commonly house an operation that is both a retailer for the average consumer and a wholesaler to the owners of smaller stands in the immediate region. These corporate-owned enterprises also serve as a distributor for many suppliers.

As is the case with many larger retail outlets, a maximum amount of the building's square footage is used for simultaneously storing and displaying the products for sale. Additionally these enterprises do not sell fireworks exclusively, but instead sell other, more-primary consumer products, such as gasoline, food items, and souvenirs. These year-round sellers have the means and need to spend more capital on the quality of the consumers' shopping experience, utilizing professional signage and providing well-lit, asphalt parking lots. The permanent nature of the buildings

FIG. 14.6. Fireworks outlet stores from a distance and up close, Tennessee. Photos by the author.

and their proximity to major highways results in bold graphics, often lighted at night, screaming at the consumer to stop in and buy. The fireworks megastores create a dilemma shared by many small towns when large, nationwide retail stores move in along the highway service roads near fireworks buildings. The local consumers reap the initial benefit of a larger scale business at the long-term expense of losing smaller mom-and-pop stores. It's a consumer-driven evolution based on the power and wealth of a few to maximize their share in the market by providing a glut of goods at a cheaper price.

RENTABLE SPACE

In order to take advantage of the inherent seasonal conditions of retail sales of fireworks, several entrepreneurs have opted to forgo the purchase of a structure to house their wares and use leaseable space in roadside strip mail outlets (fig. 14.7). By simply leasing basic, ubiquitous retail space to display and sell goods, the seller has a ready-to-go shop that needs only to be filled with merchandise. This method of selling seasonal goods is also practiced by entrepreneurs who sell Halloween costumes and Christmas decorations for only one or two months out of the entire year.

FIG. 14.7. Strip mall fireworks outlet, Alabama. Photo by Adam Bailey.

The brashness of the exteriors typically characterized by fireworks stands must be compromised within this more communal existence in a strip mall. Utilizing only the display window and an attached or removable sign as its advertising, the outlet is forced to conform its storefront appearance to that of the neighboring stores, eliminating the festive peculiarities from this particular retail venue.

SITING

As in most business ventures, "location, location, and location" is of utmost importance to a seasonal enterprise such as the selling of fireworks. Likewise, siting could be considered an even more critical issue for the fireworks outlet buildings: one can't simply pick up a five-thousand-square-foot building and move it to increase market share. While most smaller stands are sited on secondary state highways or along service roads located significantly away from the major intersections, the larger stores occupy more prominent, marketable, and expensive sites, ideally with interstate frontage and easy access.

Political and natural borders are major elements in determining ideal sites. A strategic location is just as important for the proprietor of a small stand. Many owners take their financial gamble a step further by living in travel trailers next to their fireworks stands during the sales season. The twenty-four-hour occupation of the site also creates an adequate means of securing the stand, with its lack of sturdy walls, from nighttime thieves and vandals, as well as providing a clean, well-lighted place.

SIGNAGE

As a requirement for the success of any business, getting the word out to an uninformed consumer of a superior service or product is key to financial survival (figs. 14.8 and 14.9). Given the brevity of the fireworks sales season in most regions of the country, attracting potential consumers or luring them from a competitor is paramount for a successful enterprise. Competition becomes fierce because prime locations for siting structures juxtapose business rivals within a few yards of each other. With an abundance of fireworks stands over the Fourth of July that essentially sell the same product in these prominent locations, how does an informed consumer determine where to buy? How does the fireworks entrepreneur create the illusion of a unique or better product? Clearly the quality and content of the advertising message along the roadside becomes a priority of the individual seller in his attempt to lure drivers to his particular stand.

FIG. 14.8. Fireworks signage, Missouri. Photo by the author.

One of the most economic and visible methods of advertising fireworks stands is to apply an image directly on the structure itself, either by painting on the building or placing signage on or near the stand. The use of color, the boldness and simplicity of text, and the utilization of symbolic fireworks references, such as rockets or exploding firecrackers, are common signage motifs.

FIG. 14.9. Fireworks signage, Tennessee. Photo by the author.

Large permanent retailers will expend substantial amounts of money for illuminated signs, decorative lighting, or large billboards in an effort to tempt passersby to take the next exit and buy, buy, buy. Smaller, local proprietors rely on an ingenious combination of used-car-lot and lemonade-stand approaches. Advertising is typically within easy viewing of the fireworks stand itself, although it often exists exclusively on the building. U.S. flags, lights, rockets, balloons, and foil streamers are additional gimmicks used to lure in customers. Either way, signage is big in scale and bold in design and color.

CONCLUSION

This research has been a revelation to me. I've been like a kid in a candy store. It's been an interesting experience, having the opportunity to travel the countryside with camera and notepad sandwiched between myself and my very patient wife, continually pulling over onto the highway shoulders to photograph, sketch, and describe a fireworks stand. This particular interest of mine is now well known to my family and friends. Initially most find this pursuit of documenting and describing these structures as odd at best. Paranoia strikes the heart of many a fireworks stand employee when pull my car over to capture images and take notes . . . they typically think I am a local competitor or, even worse, a member of the state fire marshal's office. However, with a few friendly words, things lighten up. Many people are profoundly curious about why I have decided to undertake this. Why not a more academic form of architecture? Again, after a few words in the form of an explanation, I have converts. Friends have been sending me images of their home state's fireworks stands. Without much convincing, they have in some ways found exactly what I have. It becomes a quest, an adventure, this looking through the windshield with a hope of discovery . . . not unlike anticipating an "ooh" or an "ah" in a grand finale.

Small towns typically rely on historic buildings to connect themselves to larger cultural histories. I have sought to reveal provocative and even poetic qualities of what have been described as the more banal and uninspired resources of our built environment. Fireworks stands, as well as bingo halls, self-service car washes, and other commercial structures are located in nearly every town in the South. Despite their rather unmemorable appearances, they cannot be denied their presence and their importance in the pursuit of daily civic life. Looking beyond the surface of their very utilitarian qualities, we see that these structures are part of the roadside landscape and they represent the values of the people that use them. This is what I try to teach my students to discover and what I attempt to elucidate in my survey of this curious contribution to the American roadside.

CHAPTER 15

VIEWPOINTS AND POINTS OF VIEW: FLORIDA'S EARLY TOURIST TOWERS

KIMBERLY D. HINDER

Perhaps the desire to stand tall and look out over the surrounding landscape is an emotion as old as time itself. When natural features such as mountains or cliffs were not available, then artificial structures were essential—even on vacation. Towers for tourists to climb as part of their vacation experience became a unique part of roadside entertainment. If that sounds strange, then remember that perhaps the best-known tourist tower in the world is the Eiffel Tower. Although such structures existed elsewhere, Florida had an unusually large number of them, in large part resulting from an economy that was heavily based on tourism and from a landscape that was flat. Like modern overlooks the towers afforded tourists an opportunity to take a break and view scenic countryside without having to traverse it. Those towers built in the 1950s and 1960s were a roadside attraction, and it is little wonder that in Florida they were concentrated on the east coast along U.S. 1, the Eastern Division of the Dixie Highway, and through central Florida along U.S. 27, much of which was the Western Division of the Dixie Highway. Florida's tourism trends have changed in the post-interstate, post-Disney era, affecting the fate of these towers.

Today, on a clear day, you can see forever. Never before, with so little physical exertion, have there been so many ways to enjoy the panoramic views once reserved for the gods.

—Judith Dupre
Skyscrapers, 1996

Since the days of the Tower of Babel and the obelisks of ancient Egypt, individuals have constructed tall objects for religious, commemorative, and defensive purposes. A well-known example, significant not only for its architectural engineering but also for the view it provided, is the Eiffel Tower. According to Joseph Harriss in *The Tallest Tower: Eiffel and the Belle Epoque,* it is "a ladder to the sky with no other fundamental purpose than allowing man to climb up and have a look around."[1] These architectural ideals were transported to the United States and were incorporated in memorials such as the Washington Monument. As the tallest masonry tower in the world, the 555-foot-tall obelisk was intended from its completion in 1884 not only to be viewed from the outside in, but also from the inside out. Like the Eiffel Tower, the view *from* the obelisk was as important as the view *of* the obelisk.[2]

FIG. 15.1. Portion of a color postcard for the Bok Singing Tower in Lake Wales. From the author's collection.

Which brings us to Florida's towers . . .

In the late nineteenth and early twentieth centuries, small towers were constructed throughout Florida, including the Clock Tower on the boardwalk at Daytona Beach and Lakeland's Citrus Tower. Most of these were simple obelisks intended to be viewed, but not to provide a view. Other towers functioned as bell towers or carillons, constructed as part of a campus or church plan. Mountain Lake Sanctuary and Singing

Tower (now Historic Bok Sanctuary),[3] located off U.S. 27 in Lake Wales, was the exception (fig. 15.1). Edward Bok constructed the carillon and surrounding gardens to thank the United States for the opportunities provided him as a Dutch immigrant. The 205-foot-tall tower, dedicated by President Calvin Coolidge on February 1, 1929, was designed to house the carillon and function as a public art exhibit combining architecture with art and music. Bok never intended for the tower to provide a view of the surrounding countryside, or even that it would be accessible to the general public. Although the tower remains closed to the public, Historic Bok Sanctuary, which is now a National Historic Landmark, continues to attract more than two hundred thousand visitors annually.[4]

FIG. 15.2. An aerial rendering of the Citrus Tower in Clermont, circa 1960, published by Koppel Color Cards and Florida Natural Color. The tower was on U.S. 27, also known as "Citrus Boulevard" and the "Orange Blossom Trail." From the author's collection.

Tourist towers in Florida, constructed to showcase a view of the landscape, developed as a result of the evolution of the state into a tourist destination. During the 1920s, tourism blossomed in Florida, attracting increasing numbers of the middle class. Despite the Depression, tourism continued to grow in the 1930s and 1940s. Propaganda promoted the state as an eternal garden with fruit and vegetables—especially citrus—available year-round. Accordingly many of the early tourist attractions, including gardens and fruit stands, lined major north-south routes that were heavily traveled by visitors during winter months. Prior to the interstate system, U.S. 1, which functioned as the Eastern Division of the Dixie Highway, and U.S. 27, which extended through central Florida to Miami, served as two of the primary north-south transportation corridors until the construction of the interstate system in the mid-1960s. The Bazaar International Trylon Tower was erected along U.S. 1 not only as a means to view the ocean and the east coast but also as part of one of the first shopping centers in the area. In contrast three other tourist towers—the Nagel Groves Observation Tower, the Citrus Tower (fig. 15.2), and the Placid Tower—appear to have been constructed primarily to invite visitors to admire the citrus groves of Florida. Developers constructed these towers on hills amid the numerous orange groves lining U.S. 27.

NAGEL GROVES OBSERVATION TOWER

FIG. 15.3. Nagel Groves Observation Tower, Winter Haven, circa-1955 postcard. From the author's collection.

The Nagel Groves tower was constructed around 1954 in the middle of the Nagel Citrusland properties between Waverly and Cypress Gardens immediately west of U.S. 27 on Highway 540 (fig. 15.3). Joseph D. Nagel, owner of several groves in the vicinity of Winter Haven and vice president of the Villa Lumber and Supply Company, purchased the land around 1952 and hired Leonard Brothers of Orlando to build an observation tower. The tower was constructed at the firm's offices in Orlando, and then Leonard Brothers delivered the tower in two pieces that were combined on site atop a concrete foundation. The one-hundred-foot steel observation tower featured 144 spiral stair steps connecting two observation platforms, one midway at the fifty-foot level and one at the top. The highest platform featured a Bausch and Lomb "viewing machine" with which to examine the "complete panorama of the symmetrical grove plantings and multitude of lakes around this densely planted citrus section."[5] Twenty-five cents purchased not only admission to the tower, but also a bag of grapefruit, while five cents allowed use of the binoculars at the top.[6]

Fourteen tons of steel were used to construct the tower, which rested on an eighteen-ton reinforced-concrete base. Guy cables attached to the structure allowed it to withstand winds up to 180 miles per hour. Nagel Groves' Citrusland Fruit Stand and Gift Shop at the base of the tower offered fruit for sale, as well as all the juice one could drink for only ten cents. Tables and chairs in front of the building allowed the tourist to sip their juice while watching the cars pass. Nagel owned two other fruit stands, the Cypress Gardens Triangle and the Nagel Groves Home Stand, both on Highway 540. The company picked fruit daily to supply these retail outlets and provided express shipment service of fruit from its own packing house.[7]

CITRUS TOWER

The Citrus Tower was the idea of Pittsburgh businessmen W. Thacker and Ferdinand Jack Toole, who retired to Clermont, a town along the central ridge of the state at the juncture of State Road 50 and U.S. 27 (fig. 15.4). When Thacker moved there in 1952, he constructed the Skyline Motel on the west side of U.S. 27 approximately one mile north of State Road 50. Along with Toole, Thacker purchased 12.5 acres on a hill directly across the highway from the motel.[8] According to Toole, "We figured this was a good stopping point. . . . It's almost the exact geographical center of the state. Cypress Gardens is forty-five miles to the south; Silver Springs is fifty-four miles to the north. We figured it was a good place for people to stop, get a cup of coffee and get a look at the scenery."[9]

FIG. 15.4. Citrus Tower, 1998. Photo by the author.

Toole sketched the first design for the tower on a piece of cardboard before passing it on to the Pennsylvania architect Thomas A. Russell.[10] Thacker and Toole formed the Florida Towers Corporation with approximately 225 stockholders from twenty-two states who invested four hundred thousand dollars in the construction of the tower. Thacker served as president and chairman of the board, while Toole was a director.[11]

The corporation initiated construction of the tower in June 1955. After thirteen months of work, it opened to the public on July 14, 1956. With 5 million pounds of concrete and 149,000 pounds of reinforcing steel, the twenty-six-foot square tower was constructed to withstand winds up to 190 miles per hour. Although the original plans specified a height of 75 feet, at completion it reached a total of 226 feet. Because the tower was located on a hill, 543 feet above sea level, visitors who took the ninety-cent ride to the top could see thirty-five miles to the horizon from enclosed and open observation decks with twenty-power telescopes. The vista encompassed portions of Lake, Orange, Polk, and Sumter counties, with the view extending to Pasco, Osceola, Seminole, and Hernando counties on a clear day. The tower provided tourists a 360-degree view of more than sixteen million orange trees.[12]

A variety of improvements to the tower and its surroundings began immediately. Soon after completion, a Schulmerich Arlington 61 carillon was installed in the tower, with Helen R. Vance, a carillonneur from Indianapolis, Indiana, performing concerts. Over time the tower also incorporated a candy shop, a citrus sales room, an ice cream parlor, a marmalade house, a glass-blowing studio, and a branch post office to draw an increasing number of tourists. At one point the Highway 27 Association of Florida operated out of the Citrus Tower Post Office. A four-hundred-foot-long map of Florida was outlined on the grounds east of the tower. Jack Toole also opened the House of Presidents Wax Museum south of the tower.[13]

PLACID TOWER

U.S. 27 also witnessed the construction of the Placid Tower (figs. 15.5 and 15.6). Earnest Oakley Hunt first dreamed of building such a tower when he moved to Orlando from Colorado in 1938. After relocating to Sebring in 1947 and visiting the rolling hills of nearby Lake Placid, he believed that he had found the perfect location for his attraction. Twelve years later Hunt, along with business associate Robert Gray, formed the Air View Corporation for the sole purpose of constructing a tourist tower.[14]

The corporation selected the site for the tower in Lake Placid because of its elevation of 150 feet above sea level. A. Wynne Howell, the Lakeland-based architect of St. Agnes Episcopal Church in Sebring, was asked to design the tower. Hunt and Gray's own construction firm, Ridge Builders, erected the structure. The tower was situated on a foundation of 520 cubic yards of concrete reinforced with 80,000 pounds of steel. Cummer Lime and Manufacturing Company provided 100,000 Ocala limestone blocks for the construction of the tower. According to the *Sebring*

FIG. 15.5. Placid Tower, circa 1961, published by Koppel Color Cards and Florida Natural Color. From the author's collection.

News, workers "were forced to use a build-as-you-go procedure. Setting up temporary elevators at each 24-foot interval, they built to the same and then raised the elevator another section. All the work was carried out by working from the inside out, under the supervision of Dayton Ormiston, the construction foreman."[15]

Once the tower walls were complete, a 2,000-pound capacity elevator carried visitors on a one-minute ride to either an enclosed observation deck at 192 feet or an open observation deck at 200 feet. Only the bravest of visitors climbed to the crow's nest, which was 225 feet above ground. Lauded as the world's tallest concrete-block structure when it opened, the tower topped off at 240 feet.[16]

FIG. 15.6. Placid Tower, 1999. Photo by the author.

Placid Tower opened January 1, 1961. The grand opening event lasted three weeks and culminated with a fireworks display from the top of the tower and more than seven thousand dollars in prizes given away to visitors. Robert Gray, who served as first vice president of the corporation, managed the tower. At its base operated a gift shop with a camera department and a snack bar. The Lake Placid Chamber of Commerce also opened an office adjacent to the gift shop. Tropical plants and sawdust walkways with rustic railings were used to landscape the grounds around the tower, and a parking lot designed to accommodate two hundred vehicles completed the tower site.[17] Nearby businesses quickly sprang up to benefit from the tower traffic. On January 8, 1961, Jack Wirick opened the Plaza Motel and Restaurant adjacent to the tower grounds. The twenty-four-unit motel boasted a pool, landscaped court, and air-conditioned rooms equipped with televisions and telephones.[18] Later the same month, a Phillips 66 service station opened at the tower, making the area a complete service center for tourists traveling U.S. 27.[19]

THE BAZAAR INTERNATIONAL AND TRYLON TOWER

Constructed in 1960, the Bazaar International and Trylon Tower were designed as a commercial retail center and tourist attraction along U.S. 1 in Riviera Beach (fig. 15.7). The complex consisted of the Bazaar International, a large two-story building that contained fifty stores and restaurants; the Bazaar Trylon Tower, a 230-foot observation structure; and a reflecting pool, gas station, and a small building intended as a day-care center. The project was the realization of developer Joseph Mass's ambitious dream of an architecturally unique plaza centered around restaurants and shops that showcased merchandise from countries all over the world.[20]

FIG. 15.7. The Bazaar International Trylon Tower, 1996. Photo by Daniel Delahaye. Courtesy of Archaeological Consultants, Inc.

The Bazaar International and Bazaar Trylon Tower opened Veterans Day, November 11, 1960, an event described by the *Palm Beach Post:* "a symbol of the market area within the breadth of

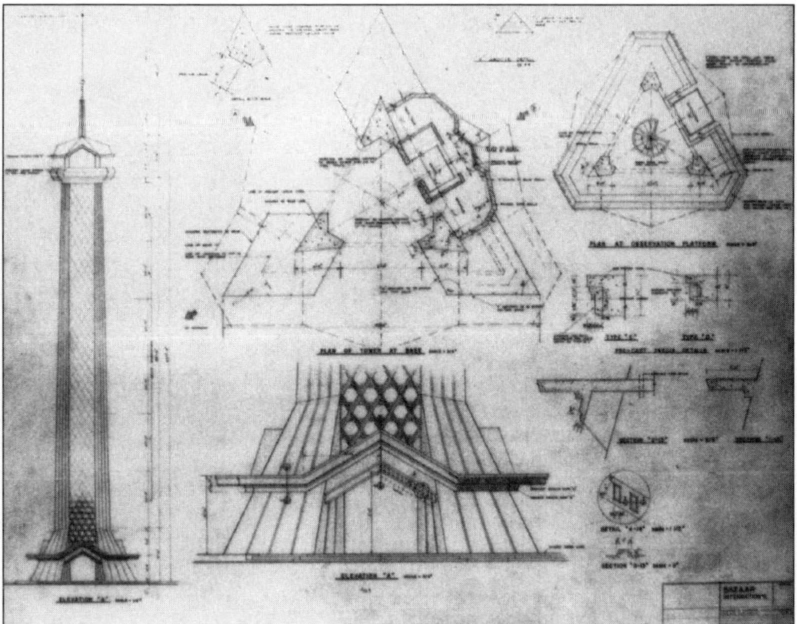

FIG. 15.8. Circa-1960 plan for the Bazaar International Trylon Tower. Courtesy Alfred Browning Parker, F.A.I.A.

the view it encompasses and its eye-catching appeal for visitors, the Bazaar Trylon Tower is expected to share the fame of Bazaar International, known as the 'world's most unusual market place' and one of Florida's newest, most exciting tourist attractions."[21]

Alfred Browning Parker, fellow of the American Institute of Architects (FAIA), designed the complex; Norman Dignum engineered it; and Butler and Oenbrink of Palm Beach constructed it (fig. 15.8). Others involved in the project were Todd Landscaping, landscape architects; E. M. Rinkor Company, concrete; Clerici Masonry and Plastering, plasterers; E. C. Peters, precast concrete forms; and Gentry Bros. Glass, glass workers.[22]

The building and tower required more than ten thousand yards of concrete and four hundred tons of steel—the largest amount of concrete and steel used in any commercial structure in Florida at the time.[23] According to Parker, the sandy soil of the large parcel originally had the consistency of sugar. Engineer Dignum compensated for the poor subsurface by "vibra-floating" the site. Probes were driven into the ground and vibrated while water was flooded into the cavity under pressure. This improved the subsurface condition sufficiently to enable construction of the Trylon complex to proceed[24] (fig. 15.9).

FIG. 15.9. The Bazaar International inner courtyard just after opening. Courtesy Alfred Browning Parker, F.A.I.A.

On its completion in 1960, the Trylon was lauded as Florida's tallest tower at 230 feet high. Its superstructure was composed of three massive triangular, concrete pillars reinforced with steel. As much steel and concrete was used below ground for the foundation as was visible above ground. Forty-two helicor pilings, each with a capacity of thirty tons of concrete, were required for the substructure; thirty-two tons of reinforcing steel were used, as well as three hundred yards of poured concrete for the base foundation. The grilled facade was made from eighteen steel-reinforced concrete panels weighing five tons each. The panels, secured to the trylon pillars, served as a stabilizing element.[25]

A glass-enclosed exterior elevator originally carried up to eight visitors on a two-and-one-half-minute ride to the observation deck at the top, for fifty cents a person. Newspaper articles claimed the elevator was "an innovation, with only two similar, but not identical elevators in the country." Three high-powered telescopes provided a broadened view from the top. Additionally a steel circular stairway for emergency use ascended the tower interior.[26] Parker intended for the tower to draw attention to the complex as an easily identifiable icon visible in the surrounding flat landscape.[27]

A reflecting pool with royal palms and the flags of various countries spanned the area between the tower and the Bazaar's main entrance, in which the image of the tower was mirrored. The lush landscaping created the likeness of a tropical oasis.[28] Hailed as the "world's most unusual market place" and a "unique Byzantine concept of an ultra-modern market place," the complex was an example of the Wrightian style, the result of Parker's emulation of the basic principals of the architect Frank Lloyd Wright. Named an AIA fellow in 1959 for his contributions to design and public service, Parker formulated his design for the Bazaar International and Trylon Tower during this period.[29] Essentially rectangular in plan with towers at each corner, the main building of the International Bazaar wrapped around an inner courtyard that attempted to "bring the outside in." Geometric details mirroring the tower and stained glass appeared throughout the building in friezes, light features, original pavers, planters, urns, chairs, and tables.[30] Immediately north of the main structure, a small building featuring the same design elements was constructed to serve as a day-care facility. The day care included an accompanying "Kiddieland" with girls' and boys' restrooms, a snack bar, and a goldfish pond. On the northeast corner of the parcel, Parker designed a service station for the Arvida Oil Company. Parker also designed a laundromat and a warehouse, but it is unknown if these were constructed.[31]

THE FATE OF THE TOWERS

Within a few years of their initial construction, Florida's tourist towers suffered financially from decreasing traffic. As a result the towers have been difficult to maintain as businesses and, in two cases, have been demolished.

After the death of Joseph D. Nagel, his widow, Mary, closed the Nagel Groves Observation Tower because of the cost of liability insurance. She leased the surrounding land before finally selling it in 1965 to Leonard Atkinson, who had worked for the Nagels at the tower and gift shop on the Citrusland properties. Mary Nagel finally sold the tower itself around 1972 to an individual who intended to incorporate it into a domestic animal farm attraction on nearby U.S. 27. The attraction never materialized, however, and the tower never reopened. It eventually decayed and was demolished.[32]

In contrast the Citrus Tower remains open, although it has had numerous owners since the early 1970s. At one time approximately 150,000 people visited the tower annually. Severe freezes destroyed many of the groves in December 1983 and January 1985, decimating the citrus industry; correspondingly visitation to the tower plummeted.[33] Many of the early exhibits, including the Bohemian Glass Blowers, succumbed to hard times in the early 1990s.

Today the Citrus Tower celebrates its fiftieth anniversary in 2006. Lilly's Citrus Tower Gift Shop and Santiago's Mexican Restaurant and Gift Shop are open at its base. The House of Presidents Wax Museum also continues to function apart from the Citrus Tower, in a building south of the attraction.[34]

The Placid Tower floundered financially by the mid-1960s and was closed periodically. Subsequent owners changed its name to the Happiness Tower, expanded the complex and gift shop, renovated the structure, and paved the area that was originally landscaped.[35] In September 1992 the tower reopened as the Placid Tower. During the first year after reopening, approximately two thousand people visited the tower monthly during the winter, with visitation dwindling to about six hundred per month during the summer.[36] The tower remains open to the public, along with the Tower Restaurant and several other commercial establishments at its base. The Plaza Motel and Restaurant now operate as separate entities as the Rosewood Senior Care Center and De Yi Chinese Restaurant, respectively. The Phillips 66 service station was demolished and replaced by a modern convenience store.

Despite the initial acclaim the Bazaar International complex failed within a few years. Situated immediately north of the West Palm Beach/Riviera Beach city limits, the site was chosen because it was believed that a bridge to the barrier islands would be constructed at the city limits, which would generate tourist traffic at the intersection. The bridge was eventually constructed several miles to the north. As a forerunner of the modern mall, the complex also lacked a major anchor store to draw in the tourists who would then visit the smaller stores and restaurants. The owners filed for bankruptcy within a few years after construction.[37]

FIG. 15.10. Line drawing of the Bok Tower. Courtesy of Historic Bok Sanctuary.

A steady decline of the surrounding neighborhood continued to plague the site. Subsequent owners installed paneling, dropped ceilings, and four large air-conditioning towers, in addition to renaming the complex the Port of Palm Beach Executive Plaza. Additionally the owners drained the reflecting pool, filled the fountain, and removed most of the flagpoles and palms.[38]

Finally, in 1992, the Port of Palm Beach purchased the plaza. Over the years the tower suffered from neglect to the point that the concrete stabilizing panels had begun to spall. Demolition was even suggested. However, in 1993, the Port of Palm Beach made the tower structurally sound once again, although the interior stairs and the glass elevator were not restored.[39] Although a request for a determination of eligibility for listing on the National Register of Historic Places was prepared for the structure, the Bazaar International Trylon Tower was demolished in March 1998 to make room for a highway overpass.[40] The main building and former day-care facility of the Bazaar International remain, in dilapidated condition, in the shadow of the overpass.

The last two decades have witnessed the decline of the historic tourist towers in Florida. Interstates 4, 75, and 95, as well as the Florida Turnpike, have drawn tourists away from the attraction-lined old roads. The obliteration of the historic environment and the citrus culture has also had a negative impact on the success of the early tourist towers. And competition from Disney World, which opened in 1971, along with that from other large-scale amusement parks and developments, has drawn visitors away from the older and smaller attractions. The concept of the tourist tower does survive and thrive in Florida. Several of the large-scale amusement parks, such as Cypress Gardens and Sea World, have incorporated observation towers into their parks. Thus the tourist tower lives.

NOTES

This essay was reprinted from *SCA Journal* 17 (Spring 1999): 4–11. Used with permission.

1. Joseph Harriss, *The Tallest Tower: Eiffel and the Belle Epoque* (Washington, DC: Regnery Gateway, 1975), 208.
2. Judith Dupre, *Skyscrapers: A History of the World's Most Famous and Important Skyscrapers* (New York: Black Dog and Leventhal Publishers, 1996), 12–13, 16–17.
3. Historic Bok Sanctuary became the official name in 2002. Previously, for many years, the site was known as Bok Tower Gardens.
4. Rebecca Spain Schwarz, "Mountain Lake Sanctuary and Singing Tower," National Historic Landmark Nomination Form, October 9, 1992. Form on file at Bureau of Historic Records, Office of Cultural and Historic Preservation, Florida Department of State, Tallahassee.
5. "Nagel Observation Tower," postcard, Nagel Groves, Winter Haven, Florida, author's collection.
6. Leonard Atkinson, interview by author, March 16, 1999, Winter Haven.
7. Ibid.; "Nagel Observation Tower," postcard, author's collection.
8. Miriam W. Johnson and Rosemary Y. Young, *Clermont: Gem of the Hills* (Tallahassee: Rose Printing, 1984), 253.
9. Jan Glidewell, "On Top of U.S. 27," *St. Petersburg Times,* July 26, 1987, F1, F11.
10. Greg Homan, owner of Citrus Tower, interview by author, March 8, 1999.

11. Glidewell, "On Top of U.S. 27," July 26, 1987.
12. Ibid.
13. Johnson and Young, *Clermont*, 253; Don Dornbrook, "Clermonts Tribute to an Industry," *Leesburg/Commercial*, September 11, 1986, vertical file, Cooper Memorial Library.
14. "Large Crowds View Placid Tower Sunday," *Sebring News*, January 5, 1961, 1B.
15. Ibid.
16. Ibid.
17. Ibid.
18. "New Tower Motel and Restaurant Open Sunday," *Sebring News*, January 12, 1961, 2B; "Placid Tower Surroundings Are Beautiful," *Sebring News*, March 16, 1961 (clipping, Sebring Historical Society Archives, Sebring, FL).
19. "New Service Station at Placid Tower," *Sebring News*, January 19, 1961 (clipping, Sebring Historical Society Archives).
20. "Bazaar Trylon Tower Ceremonies Today," *Palm Beach Post*, November 11, 1960, 15–17.
21. Ibid.
22. Ibid.; Alfred Browning Parker, FAIA, interview by Daniel Delahaye and author, January 20, 1997.
23. "Bazaar Trylon Tower Ceremonies Today," 15–17.
24. Parker interview, January 20, 1997.
25. "Bazaar Trylon Tower Ceremonies Today," 15–17.
26. "Bazaar Trylon Tower Ceremonies Today," 15–17.
27. Parker interview, January 20, 1997.
28. "Bazaar Trylon Tower Ceremonies Today," 15–17; "The Many Lives of a Plaza," *Gateway: Magazine of the Port of Palm Beach* 16 (October/November/December 1993): 11–12.
29. Parker interview, January 20, 1997.
30. "Bazaar Trylon Tower Ceremonies Today," 15–17.
31. Alfred Browning Parker, original drawings of the Bazaar International, personal collection of Alfred Browning Parker, 1959–60.
32. Atkinson interview, 1999; Polk County, Florida, property appraiser, tax parcel records.
33. Adam Yeomans, "A Citrus Tower Is Sold," *Orlando Sentinel*, November 27, 1986, B1 (clipping, vertical file, Cooper Memorial Library).
34. Homan interview, March 8, 1999.
35. "Towers Gives Visitors a View of Countryside," *Tampa Tribune*, October 26, 1992.
36. "The Tower: Lake Placid's Landmark Gets a Facelift," *Sebring News-Sun*, July 11, 1993.
37. "The Many Lives of a Plaza," 11–12.
38. "Office Complex Will Occupy Old International Bazaar," *Palm Beach Post*, April 24, 1973; "Executive Plaza Offices Popular," *Palm Beach Times*, November 29, 1973.
39. "Port OKs $123,000 to Save International Bazaar Tower," *Palm Beach Post*, December 16, 1993; "Rusting Riviera Tower Gets Face Lift," *Palm Beach Post*, March 12, 1994; "The Many Lives of a Plaza," 11–12.
40. Daniel Delahaye and Kimberly Hinder, "Bazaar International and Trylon Tower," National Register Nomination Form (request for a Determination of Eligibility), August 26, 1996, updated January 1997. Form on file at Archaeological Consultants Inc., Sarasota, FL.

CHAPTER 16

SEEING THE SCENIC UPLAND SOUTH: MOTHER NATURE AND THE MORPHOLOGY OF TOURIST LANDSCAPES

KEVIN J. PATRICK

One of the reasons tourists head to the South is to enjoy the natural beauty of the region. At the same time, tourists and entrepreneurs stimulate great changes in the landscape, which often result in a conflict between Mother Nature and the tourism industry. The exquisite landscape is often described in near religious terms. How people affect this landscape is rarely described in such glowing words. Travel had been less frequent and slow, but then the railroads began to transport more people faster, and then, in turn, the automobile superceded the trains and brought even more travelers. Rails were site specific: there were set routes and times to travel, and when riders got to a destination, they were often restricted to horses and carriages or foot traffic. Automobiles and roads allowed more choices. This freedom to come and go whenever and wherever benefits both business and recreational travelers, but everyone might not consider the impact on the land a benefit.

Once a wanderer, seeking the Pisgah heights, was cheated in trade for food. In writing of the experience he said, "It matters little. In Vagabondia one does not haggle over the price of this or that; on the open road money is a small thing indeed, and it is better to be cheated out of the change than to miss a noble prospect of far-off mountains or a fir tree by the roadside."

—John T. Farris

Seeing the Sunny South, 1921

TWO PATHS DIVERGING IN A YELLOW WOOD

U.S. tourism has always contained a paradoxical duality. In part the goal of vacation is to escape from the hectic, urbane, work-a-day world and find a secluded, natural setting in which to relax and recharge. On the other hand, there is the desire to do something exciting and out of the ordinary, to be stimulated with good food and unique diversion while being surrounded by all the comforts of home. U.S. tourist landscapes mirror this duality, especially the ones constructed around scenic wonders where majestic mountain peaks compete with amusement parks and wax museums. The Upland South, including southern Appalachia, the interior low plateaus of Kentucky and Tennessee, and the Ozarks Mountains of Arkansas, are replete with such landscapes. Despite the paradox southern highland vacationers have come to accept that it is normal for the kind of scenic splendor exhibited by the Great Smokies to be indivisibly tethered to the type and magnitude of contrived, tourist diversions found in nearby Gatlinburg and Pigeon Forge, Tennessee. How did such places come to coexist? How can these landscapes be interpreted, and what kind of meaning can be derived from them?

The complex morphology of nature-related tourist landscapes in the Upland South reflects two and one-half centuries of tourism, which can be subdivided into just four periods defined by the prevailing interpretation of nature. In each period the primary aesthetic is apparent not only in the landscape, but in the dominant visual medium of the day. Landscape painting played a major role in disseminating the landscape aesthetic during the Victorian era, which was later picked up by photography and picture postcards, and finally by the electronic media of today. An examination of contemporary visual media provides insight on the meaning of tourist landscapes. Their appearance, form, and function, however, are largely the result of the dominant mode of transportation.

These four landscape aesthetics discussed below are nature as wilderness cathedral, nature under control, nature as road-sideshow, and nature as recreation. The landscape aesthetic of each successive period is not obliterated as much as a more-dominant perspective overlies it. Elements of each are still discernible. The four aesthetic periods chart the evolution of landscape from the sublime and picturesque, through its redefinition as a utilitarian resource, to its interpretation as tourist attraction and recreation park. In order to really see the scenic Upland South, the journey must begin back in time when North America's natural landscape was a sublime, howling wilderness.

NATURE AS WILDERNESS CATHEDRAL

Once, all of Appalachia was wilderness, and that wilderness was revered as sublime, the direct manifestation of God's handiwork: immutable, insurmountable, and horrifically grand. After the Atlantic forests had been rolled back by farmsteads, however, the scenic wonders near at hand came to be celebrated as wilderness cathedrals. The savage, fearsome spectacle of divine sublimity receded to include a human component, represented by a pensive voyeur on a craggy ledge, or an ashen wisp of cabin smoke surrounded by God's purple mountain majesty.

Like a church, the wilderness cathedral could be entered for periods of self-reflection, spiritual renewal, or communion with the Almighty. The appropriate sentiments were to be emotionally inspired by the natural surroundings.[1] Ritualistic renderings in paint or verse heightened the experience while providing a record that could be disseminated among the uninitiated. Landscape painters of New York's Hudson River School, such as Thomas Cole and Thomas Moran and nature writers such as Henry David Thoreau and Ralph Waldo Emerson, rose to take their positions as the masters of the new landscape aesthetic.

At the dawning of the nineteenth century, the wilderness cathedral was just beyond the inhabited fringe astride the backbone of the Appalachian Mountains. Southern Appalachia's earliest tourists were these seekers of the sublime, and few natural wonders epitomized the notion of the wilderness cathedral more than Natural Bridge, Virginia. Located at the southern end of the Shenandoah Valley, this forty-foot thick, dolomitic arch spans a two-hundred-foot deep gorge as the last remnant of a cavern roof that had encased Cedar Creek eons ago.[2]

Describing Natural Bridge, Thomas Jefferson wrote, "It is impossible for the emotions arising from the sublime to be felt beyond what they are here; so beautiful an arch, so elevated, so light, springing as it were up to heaven. The rapture of the spectator is really indescribable."[3] Indescribable, but not unobtainable. In 1774 Jefferson bought the bridge and 157 acres from King George III for twenty shillings. After some time he erected a small building near the arch, and entrusted his slave Patrick Henry with the job of caretaker. Patrick's job was to keep one or two rooms of this building in readiness to receive Jefferson or his friends whenever they might come. At Natural Bridge sublimity had been sold as private property to be overseen by a guide who maintained accommodations for visiting guests. With the United States in its incipiency, the South had its first scenic wonder tourist attraction.

The massive rock bridge, which was frequently compared to the better-known Niagara Falls, was the epitome of the wilderness cathedral. One visitor to Natural

Bridge, Samuel Kercheval, wrote a poem about the site based on divine imagery for his 1833 book, *History of the Valley of Virginia.* Natural Bridge has retained its romantic sense of the sublime. Modern conveniences, such as electric lighting, have not subsumed these initial environmental perceptions but have been employed in a deliberate attempt to enhance them. Since 1927 visitors have been treated to a nightly spectacle of illumination titled the *Drama of Creation,* during which visitors observe the "soft glowing of diffused light moving as the music moves—mingle in an unforgettable, soul stirring andante of all time—nature's own symphony out under the stars where one may walk alone with God."[4]

The antebellum South's only other scenic attraction to rival the notoriety of Natural Bridge was Mammoth Cave, Kentucky. Like Natural Bridge, Mammoth Cave was located just off the beaten path at the wooded, upland edge of an expanding agricultural frontier. Halfway between Kentucky's Bluegrass Region and the Nashville Basin of Tennessee, Mammoth Cave's mighty maw opens on a wooded glen in the Green River valley. Hunters came upon the cave's incomparable opening in the 1790s, and as early as 1816 commercial tours were being conducted. By then the western frontier was beyond the Mississippi. Kentucky and Tennessee were states, and the South's booming agricultural economy was actively sowing the seeds of a new class of nature seekers, the aristocratic tourist.

Mammoth Cave's wilderness aesthetic was sometimes portrayed as an image of nature akin to a Gothic horror story. Alexander Clark Bullit's *Rambles in the Mammoth Cave in the Year 1844* describes the experience of a typical tourist party being led to the cave as being a gloomy and dark spectacle.

To be sure, the visitors were not as much terrified as they were entertained by the suggestion of terror. The imagery of impending doom was underscored by the names of the formations in this fanciful underworld: Gothic Avenue, Devil's Arm-Chair, Lover's Leap, Wandering Willie's Spring, Giant's Coffin, Deserted Chambers, Bottomless Pit, Dead Sea, Lake Lethe, and of course, the River Styx.

Following a tradition pioneered by European royalty, the U.S. leisure class also engaged in the practice of "taking the waters."[5] This entailed seasonal pilgrimages for prolonged stays at mountain spas in order to be physically and spiritually renewed by the healing power of nature. The key feature of the spas was the water that issued forth from various mineral and thermal springs. The assumed healing powers of the water, which was ingested and bathed in as part of a daily regimen, was replete with religious symbolism of purity, baptism, and rebirth that complemented the aesthetic appreciation of the wilderness cathedral.

Although the mountain resorts offered its patrons views of sublime wilderness, their landscape aesthetic tended toward the picturesque, containing more intimate vistas that were pleasingly varied, rough, and irregular and that could include select elements of human artifice as long as they enhanced the romantic appeal of the scenery. These could take the form of a winding path, a rustic gazebo or bridge, a fence row, a villa partially obscured by vegetation, or a mowed greensward highlighted with a rocky outcropping, copse of trees, pond, or even a farmer's field or pasture.

Similar to that of the manor house of a southern plantation, the morphology of the mountain spa centers on the grand hotel. The hotel was typically set behind a greensward or an allée of trees, with a cluster of cottages nearby. Private carriage service connected many resorts to the nearest town or tavern on the main wagon road. These gateways not only funneled guests to the resort, but also acted as service centers, providing supplies and access to the outside world via postal routes and commercial stagecoach lines. Once the resort grounds were reached, patrons were catered to while being enveloped in an oasis of class-conscious civility that isolated them from the mundane everyday world around them.

One of the most essential features of the grand hotel was the veranda.[6] This full-sized porch, frequently wrapped around the building on multiple floors, was the physical adaptation to an appreciation of nature that was overwhelmingly visual. The hotel was set to provide the veranda, as the main, public observation gallery, with the best scenic vistas. Comfortably seated in a white wicker chair between formal hotel and the outdoors, a patron, sketchbook in hand and protected from the weather, could view the wilderness cathedral as a picture, literally framed by the colonnade supporting the porch. Gazebos, located about the grounds and providing other vistas in a more private, intimate setting, were miniature extensions of the hotel's veranda.[7] As the hotel's primary informal public space, the veranda was the daytime social nexus where self conscious guests observed each other as much as the scenery.

Resorts located near well-traveled wagon roads had an obvious advantage over "more sublime" sites deeper into the mountains. White Sulphur Springs, Virginia (now West Virginia), prospered along the James River and Kanawha Turnpike, Tidewater Virginia's main road through the Appalachians to the Ohio Valley.[8] The first inn at White Sulphur Springs opened in 1780. The resort was enlarged and improved in 1858 with the construction of a hotel known as the Greenbrier. The Greenbrier was the queen mother in a colorful galaxy of mid-nineteenth-century springs that included Sweet Chalybeate, Blue Sulphur, Red Sulphur, Gray Sulphur, and Green Sulphur.[9] Thirty miles northeast, another preeminent resort, the Homestead, opened in 1766. Both the Greenbrier and the Homestead are still in business.

Settled a generation after the southern Appalachians, the Ozark Mountains of Arkansas experienced a similar geographic development. Located at the edge of a prosperous agricultural region that included the plantation-rich lower Mississippi

Valley, the Ozarks were an upland pocket of marginal farms and natural wonders. Surrounded by the Zig Zag Mountains of the eastern Ouachita Range, the "Valley of the Vapors" contained forty-seven hot springs gushing forth a million gallons of water a day at an average temperature of 143 degrees Fahrenheit.[10] Although mountain-bound, this retreat was not far from the main lowland road between Tennessee and Texas, which brought cure-seeking clients at an early date to establish what became Hot Springs, Arkansas.

So significant was the region that in 1832 the federal government set aside four sections containing Hot Springs as a reservation. This was an unprecedented act but consistent with the emerging U.S. culture of wilderness interpretation. By that date the Hudson River School vocabulary used to appreciate natural landscapes was well established as was the mountain spa setting in which these landscapes were to be appreciated. When the tourist frontier pushed into the Ozarks, travelers knew what they were seeking.

Although the mountain resorts were established in the romantic idiom of the sublime and picturesque, their social function as upper-class meeting grounds came to dominate. In the shift the visual landscape aesthetic gave way to diversion. The image of nature as a revered, wilderness cathedral, carved by the hand of the Almighty, dwelt in by the noble savage, faded. To many resort patrons, the natural setting became an elaborate backdrop to gambling, horseback riding, boating, and socializing.

NATURE UNDER CONTROL

Cole's *Expulsion from the Garden of Eden* depicts a fearful Adam and Eve being sent from the holy picturesque into the terrifying sublimity of God's wilderness cathedral.[11] Lest one forget, Adam and Eve do not perish in the wilderness, but thrive. Their punishment is that they have to work for a living, and from this "fortunate fall" all people are availed to employ their own ingenuity to carve beauty from the wilderness.[12]

With their manifest destiny to bring civilization to the North American Land of Nod, Anglo-American settlers and speculators traded awe-inspiring sublimity for pragmatic utilitarianism. Many of the Upland South's scenic wonders were put to practical use soon after discovery. During the Revolutionary War, Virginia's Natural Bridge was used as a makeshift shot tower to manufacture bullets (fig. 16.1). A cave located just upstream was mined for saltpeter to make gunpowder. The bridge was also put to use as a convenient carriageway across the gorge, a role it still fulfills by bearing the traffic on U.S. 11. During the War of 1812, Mammoth Cave was turned into a saltpeter mine. The abandoned miner's cabins were used as the core structures to the first Mammoth Cave Hotel.[13] The U.S. landscape transition from sublime, through picturesque to the pastoral beautiful, and eventually to the urbane was accepted as the natural course of civilization, aided and abetted by the landscape painter's most powerful symbol of progress, the railroad. The railroad entered the picturesque wilderness cathedral and converted nature into culture on an ever-increasing scale.[14]

As the main protagonist for bringing nature under control, the railroad was also the vehicle that radically altered the Upland South's tourist landscape. Perhaps because of their owners' sense of environmental entitlement, railroads pushed through the varied landscapes to extract the timber, coal, and other resources that kept the railroad industry solvent and expanding. The railroad opened more highland areas to an increasing number of nature tourists, while sowing progress in its wake, as evidenced by the forest of stumps it left behind.

FIG. 16.1. Natural Bridge, Virginia, is the Upland South's premier example of wilderness sublime. From the author's collection.

Although the South's agricultural elite were decimated by the ravages of Civil War and Reconstruction, rail access to wealthy industrial centers, especially in the North, stimulated a rapid increase in both the number and patronage of mountain resorts in the Upland South. The mountains of Virginia were breached in 1873 by the Chesapeake and Ohio (C&O) Railroad, which came to control the James River route from Tidewater Virginia, past Natural Bridge and White Sulphur Springs, to the coalfields of the Kanawha Valley and beyond. A branch line was extended to Hot Springs, Virginia, and soon railroad executives, northern industrialists and financiers, and other major players began arriving at the springs in private railcars. In 1910 the C&O Railroad bought the Greenbrier.

Farther south a string of religious resorts and retreats followed the Southern Railway into the mountains of North Carolina. The Southern Baptists camped at

Ridgecrest, the Presbyterians established Montreat, and Lake Junaluska became the "Summer Capital of American Methodism." Hot Springs, North Carolina, was developed as a thermal spa after the 1882 trans-Appalachian rail link between Asheville, North Carolina, and Knoxville, Tennessee, was opened. Then in 1908 the Carolina, Clinchfield, and Ohio drove another line across the mountains via Spruce Pine, North Carolina, to bring Appalachian coal to the industrializing Carolina Piedmont. Almost as a matter of course, the railroad built Altapass, a mountain resort at the crest of the Blue Ridge.

With the railroads being driven by capitalist muscle, what had been the tentative control of the United States over nature became a full-blown conquest. Every tie and rail thrown down measured the cadence that beat down the walls of wilderness, destroying the limitations posed by time and space. The sudden appearance of the machine in the garden brought with it the image of domineering industrialization as a counterforce to Jeffersonian pastoralism. At first the pastoral artists attempted to accommodate the new machine, but they eventually left its progress to be recorded and celebrated by another industrial age artifice: photography.[15] Complementary technologies, such as inexpensive lithography and the adoption of the penny postcard, disseminated the conquest of nature to the masses.

Two hundred miles southwest of Virginia's Natural Bridge is the fifteen-hundred-foot Natural Tunnel. Although the two rock underpasses are located in the same physiographic province and have been eroded from the same limestone, their fame and popular depiction differed widely. Despite William Jennings Bryan's proclamation that Natural Tunnel was the "Eighth Wonder of the World," its notoriety never reached that of its more bridgelike twin. Its claim to fame, celebrated in countless postcards, came after 1890, when the South Atlantic and Ohio Railroad laid tracks through the tunnel to gain access to Appalachian coalfields. In nearly every portrayal of Natural Tunnel, the tracks are depicted not as a despoiler of nature's grandeur, but as a clever use of an otherwise useless hole.

Likewise the gorge of the Kentucky River southwest of Lexington is a mere picturesque backdrop for the Cincinnati, New Orleans, and Texas Pacific's High Bridge. A 1937 Curt Teich postcard celebrated its wilderness-waffling dimensions by defining it as the "highest bridge spanning a navigable stream on the American continent. . . . The most massive structure of its length and breadth known to the civilized world. Height 315 feet. Length 1,223 feet."[16] Equally astounding was the Atlanta, Knoxville, and Northern railroad's 1898 Loop over Bald Mountain. To gain altitude while maintaining a low grade, the railroad wrapped a spiral track around a knob overlooking the Ocoee River. Here a postcard-type view presents a cut and shaved mountaintop as a challenging obstacle subdued by human ingenuity.[17]

The railroads brought the comfort and connectivity of urbanity to now-tamed wilderness settings. Added to the ensemble of hotel, cottage, and pavilion was the depot and the branch line. In many cases the tracks were laid right to the threshold of the resort. The relatively high density environment was necessary to accommodate visitors who were foot-bound after stepping from the train.

Beyond the structures the natural setting was more pastoral, less sublime. Doctors had begun to recommend vigorous exercise in addition to mineral-laden spring water as the road to health, stimulating a demand for recreation and its associated landscapes. Bridle paths crisscrossed wooded areas, boats dotted picturesque lakes, and tennis courts appeared among the pines. Ranking supreme among them was the golf course, a collection of sweeping greenswards and roughs manipulated in the beautiful idiom to mock the Scottish countryside.

Resort hotels capitalizing on mountain scenery and mineral waters opened all across the South. The nine-mile Mammoth Cave Railroad opened to the hotel's doorstep in 1886. Nearby Park City, Cave City, and Horse Cave, with their key locations on the railroad, expanded their economic bases with numerous commercial buildings and hotels. In Arkansas the entire spring-resorts industry blossomed as a result of the railroad. In 1876, when the town of Hot Springs was separated from the U.S. reservation, major investments in hotels and bathhouses began. A little more than sixty years later, Hot Springs boasted seventy-two hotels, its well-known Bathhouse Row, the Army and Navy General Hospital, and a thirty-day horse-racing season.[18] In Virginia in the 1880s the railroad promoter Col. Henry C. Parsons took over Natural Bridge, at first enlarging an existing small hotel and later building a larger one. Parsons enjoyed the proceeds of a time when "the railroads [ran] excursion trains, bringing people by the hundreds to see [the] wonder; when all the available transportation facilities of the management—stage coaches, surries, station wagons, carriages—would be called into play, and sometimes fall short of accommodating the crowds."[19]

Rail-based transportation systems radiating out from major upland population centers allowed for the tourist development of scenic wonders and picturesque settings on local mountain tops. Although out-of-town tourists were courted, the mountaintop parks and resorts were particularly dependent on day-trippers from the nearby city. Once again, nature was well under control, and frequently up-staged by the feats of engineering required to reach the summit. Picturesque settings were garnish for elaborate hotels known as much for dancing, drinking, and gambling as for the natural environment. Even the spectacular views from obligatory overlooks and observation towers were less a panorama of sublime wilderness than they were great overviews of the city at the base of the mountain.

Perhaps the South's best-known urban hilltop retreat was Lookout Mountain, a large, cliff-rimmed plateau that rose above Chattanooga, Tennessee. From 1886 to 1895 four rail-based conveyances were constructed to ridge-crest hotels.[20] The impossibly steep bluffs were breached by two inclined railways—engineering wonders of their day—the Chattanooga and Lookout Mountain Railroad, which served the 1888 Lookout Inn Hotel, and a narrow-gauge line built from Incline Number 1 (opened in 1886) around the rim of the plateau to the Point Hotel. Within five years of its 1895 opening, the second Lookout Mountain Incline, advertised as the world's steepest, had driven the other conveyances off the mountain. At the base of the inclines, St. Elmo emerged as the gateway community between Chattanooga and Lookout Mountain. Bound by rails, covered with summer cottages, upper-income housing, resort hotels, and, after 1890, the romantically landscaped Civil War battlefield memorial at Point Park, Lookout Mountain's natural locale was subsumed by human artifice.

In the late 1920s Curt Teich and Company published a twenty-view postcard folder of Eureka Springs, a late-nineteenth-century resort town in Arkansas, that summarized the Upland South's nature tourism at the end of the railroad's era of conquest. The twenty souvenir images showed four picturesque or pastoral landscape views, three views of the heavily landscaped springs, two diversions in the form of Eureka Springs Golf Course and the Lake Lucerne Amusement Park, and three views of the town and hotels. In addition to these urbane images, there are no less than five different views of the railroad, illustrating tunnels, bridges, and smoke-plumed trains steaming through massive cuts. The remaining views in the folder also depict conveyance, but of a different means. Three images are devoted to road cuts and bridges on U.S. 62, the path of a new machine in the garden following in the footsteps of the railroad.

NATURE AS ROAD SIDESHOW

When the automobile became a popular form of transportation in the early twentieth century, national civility rode the rails. The railroads may have bested time, space, and nature, but many miles of road were still wrapped in an anti-auto wilderness. For *Seeing the Sunny South* (1921), John Faris piloted an automobile throughout the Southeast. When it came to the high mountains of the southern Appalachians, however, Faris abandoned the open road for the reliability of the rails.

Despite the challenge auto enthusiasts had nothing but faith in their new machine-age marvel. The wheels of the Good Roads Movement were already turning to subdue even the remotest regions of the South. Politically strong progressives combined resources earmarked for farm-to-market roads to improve a few long-distance tourist highways, specifically those coming out of the North.[21]

The most effective Good Roads effort in the South was that of the Dixie Highway Association. Founded April 3, 1915, in Chattanooga, Tennessee, the Dixie Highway angled through the Midwest and across the South from Saulte Ste. Marie, Michigan, to Miami, Florida. It was the dream of Indianapolis headlight manufac-

FIG. 16.2. Literally following in the footsteps of the parallel Louisville and Nashville Railroad, the Dixie Highway (U.S. 25) carried tourists from the Midwest into the southern Appalachians, beginning with the crossing of the Cumberland Mountains at Jellico, Tennessee. From the author's collection.

turer Carl Fisher, who had successfully organized the Lincoln Highway Association to build the nation's first transcontinental highway two years before. Contrary to the owners of the railroads, who built resorts to encourage ridership, Carl Fisher promoted the road in order to build a resort. Fisher had invested heavily in Miami Beach waterfront property and needed the highway to promote this real estate extravaganza as a destination to winter weather-weary Midwest motorists.

Other named highways followed. Chief among them, concerning access to the Upland South, was the Robert E. Lee Highway, extending from New York to New Orleans via Washington, the Shenandoah Valley, and Birmingham, Alabama. Crossing at Chattanooga, the Dixie and Lee highways formed a giant "X" across the South that opened the scenic wonders of southern Appalachia to auto tourism. This motor age conquest of nature was all but assured when the federal government began to back the building of public roads with the Road Act of 1916 and the Federal Highway Act of 1921.

As the early railroads had inherited the landscape aesthetic of the preexisting Hudson River School, the automobile inherited the aesthetic of the railroad, that of dominance over nature. Postcard photographers who once rode the rails now took to the roads, recording the conquest with countless views of stark road cuts, soaring bridges, and miles of fresh pavement. The similarity between pictured roadscape and railscape was not incidental. Many highway engineers learned their trade on the railroad and simply transferred their design penchant for tangents, arcs, and level roadbeds to the new medium.

As reducers of natural barriers, roads not only followed in the footsteps of the railroads, they began to replace them. Increased auto usage and the extension of hard roads throughout the 1920s would leave the Mammoth Cave Railroad too weak to survive the Great Depression. It closed in 1931.[22] In Roanoke businessman W. P. Henritze pulled a page from the railroad's design book in building Prospect Road to the top of Mill Mountain. The concrete road, including tight switchbacks and the famed "figure-8" loop, opened in 1927 and proved to be so popular that the Mill Mountain Incline went bankrupt within two years and was dismantled in 1930. Another railroad-inspired loop was constructed in the 1930s on the Newfound Gap Highway over the Great Smoky Mountains in anticipation of this route becoming the main road through a new national park.

Down in Chattanooga the automobile was also changing the face of Lookout Mountain (fig. 16.3). Tennessee's first federal aid road project, the Will Cummings Highway, was carved out of the mountain's north face at Jonas Bluff in 1918 and extended to the mountain top as the Scenic Highway. In 1931 the old Whiteside Turnpike was rebuilt as the Ochs Highway, providing free vehicular access to Point Park from St. Elmo.

At the same time that the road and motorcar were becoming more reliable, they were also becoming less expensive, spreading the joys of individual freedom of movement to a wider spectrum. As an increasing number of people came to en-

FIG. 16.3. Referring to the guest register at the War Museum in Point Park, the size of the "Big Book" on Lookout Mountain suggests the relative significance of human artifice and diversion in nature-related tourist landscapes. From the author's collection.

counter nature as framed by the windshield of a car, a new landscape aesthetic began to emerge alongside that of conquest. In a hurry to get to their destination, windshield tourists experienced nature as a sideshow to be taken in during short rest stops on otherwise long trips. The scenic wonder, be it an overlook vista, a strange rock formation, a waterfall, or a cave, would present itself in the view field of the passing motorists and entice them to stop. The natural curiosity was inspected, possibly recorded with a photograph, and then left in the dust. As a leisure-time activity, auto touring was akin to a day at the beach or an amusement park, and the sights along the way were in many ways analogous to the contrived attractions of the midway.

Vividly colored linen postcards, two parts photo and one part picturesque landscape painting, implied that the southern highland roadside was filled to the brim with Profile Rocks, Chimney Rocks, Pinnacle Rocks, and Lovers Leaps, each with their own tragic love story ending in the plummeting death of a young Native American maiden. Somewhat less than sublime, the noteworthy attraction of North Carolina's Bridal Veil Falls was the routing of U.S. 64 behind the cascade, creating the watery equivalent of the drive-thru redwood tree.

It did not take roadside entrepreneurs long to realize that tourists prefer to combine their sightseeing stops with the money-spending necessities of travel, such as gas, food, lodging, diversion, and souvenirs. Filling stations, cafes, and cabin courts staked out the ridge tops, such as Mount Gayler on U.S. 71 in Arkansas, Clinch Mountain on U.S. 25E in Tennessee, and Massanutten Summit on the Lee Highway in Virginia, where overheated engines forced tourists to stop and take in the view . . . and possibly a Coke . . . and a few gallons of gas . . . and maybe some souvenirs. . . .

Unlike the clustered morphology of the railroad resort, the automobile created a linear landscape stretched along the berm of the road. Many well-traveled highways averaged one filling station per mile, and the entire auto-dependent commercial roadside only increased its density over time, especially at the approaches to towns or attractions.[23] The trackside hotels in Mammoth Cave's gateway communities of Park City and Horse Cave fell out of favor under the competition of roadside motor courts operating along the Dixie Highway at the edge of town. This included Frank Redford's Wigwam Village, which opened in 1933 as a whimsical extension of road-sideshow attractions that contrasted with the staid formality of the resort hotels and their more elitist landscape aesthetic.

Larger natural attractions such as caves could garner an admission charge to support a mini–roadside empire heralded by a barrage of billboards scattered over hundreds of miles to draw in the curious (fig. 16.4). As early as 1910, Virginia's Luray Caverns was already attracting auto tourists. Bound for Charleston, South

FIG. 16.4. Sitting in the shadow of Cumberland Gap, Cudjo's Cave was part of the early roadside show along U.S. 25–E. From the author's collection.

Carolina, Connecticut native Seymour Cunningham, deviated miles off the National Highway over rough roads to see the cave's wonders.[24] With the construction of passable roads throughout the Upland South in the 1920s, previously remote and isolated caves were developed and thrown open to the money-toting, motoring public. By the 1920s seventeen different caves were running tours in the vicinity of Mammoth Cave, Kentucky.[25]

The commercial cave was the perfect representation of nature as a road sideshow. As timeless and somewhat mysterious natural wonders, caves carried a hint of fearful sublimity. Gone, however, was the sense of wilderness and the discomfort it implied. In its stead was the grotesque picturesque seen as vignettes along a winding path through a surrealistic underworld garden. Fanciful formations continued to be given romantic names in the Gothic tradition or illustrative of antiquity, legend, and lore. Visitors to Luray Caverns still follow the Elfin Ramble past ethereal images of Pluto's Ghost en route to Hades and Giants Hall. Owners used paths, stair steps, bridges, and especially electrical lighting to enhance the picturesque qualities of their caves, walking the line between scenic attraction and human contrivance, between displaying nature and displaying a freak of nature.

In some cases the human contrivance used to make these alien worlds accessible became attractions in their own right. The spectacular Ruby Falls, a 145-foot waterfall inside Lookout Mountain, was described on a 1930s postcard as "worthy of the Master's Hand alone."[26] Nonetheless visitors reached the sublime, subterranean spectacle by first descending 260 feet in an elevator shaft hewn out of solid rock in 1929.

This was a technical response to a solution to an earlier problem. During the utilitarian age of conquest, the Southern Railway had sealed off the natural entrance to Lookout Mountain Caverns in the building of a tunnel through the ridge.

Twenty years later a 258-foot elevator shaft was blasted out of granite, using eight tons of dynamite, to take tourists to the top of North Carolina's Chimney Rock in a mere thirty-two seconds.[27] With its meandering, picturesque paths, whimsically-named rock formations, and staged vistas, Chimney Rock, North Carolina, was like a cave turned inside out. Instead of vantage points revealing metaphorical cave scenery, such as frozen waterfalls, dream lakes, and calcite formations that looked like natural landscapes, Chimney Rock's vistas presented the tourist with actual scenery, including Hickory Nut Falls, Lake Lure, and views of forested mountains that rolled off to the horizon. Chimney Rock was opened to the public by Jerome Freeman in 1885, but its real development as a scenic tourist attraction did not occur until after it was purchased by Lucius Morse in 1902. Chimney Rock overlooked the rugged Hickory Nut Gorge, the route not taken by the railroads to cross the Blue Ridge Mountains. The nearest tourist railheads were in Hendersonville and Rutherfordton, fifteen and twenty miles away. Automobile access was necessary to turn this eons-old, rocky pinnacle into a road sideshow. Accordingly Morse's first improvement to Chimney Rock was the building of a three-mile-long road to its base, which opened in 1916 on Independence Day, the height of the summer automobiling season.

Lookout Mountain was the South's undisputed king of the natural roadside attractions. The railroads claimed the mountain as a ridge-top resort for Chattanooga elite, but the automobile opened it to the masses, spreading its fame far afield. Like an auto tourist beacon, Lookout Mountain stood near the intersection formed by the Dixie and Lee highways, and its road sideshow attractions were telecommunicated up and down both routes in the form of brightly painted billboards and barn roofs proclaiming, "Ruby Falls!" and "See Rock City."

The story of Rock City Gardens illustrates the transition between the tourist landscape aesthetic of the railroad resort and that of the automobile's road sideshow. In 1924 Garnet Carter bought six hundred acres of Lookout Mountain in Georgia that included locally famous rock formations. Carter's property was not far from where the Chattanooga and Lookout Mountain Railroad reached the top of the plateau through a cleft in the bluff. Within a year he had built his own rustic, Tudor-style inn among the rock formations and had the grounds landscaped in the Olmsted tradition. In consort with the picturesque aesthetic, Carter drew on the romantic appeal of the fanciful setting, as well as his wife Frieda's penchant for fairy tales, and named the resort, Fairyland. In addition to the Fairyland Inn, the summer resort included a cluster of cottages Frieda had named Mother Goose Village.[28]

Typical for resorts of the day, Fairyland had tennis courts and a swimming pool; it was also adjacent to the Lookout Mountain Golf Course, opened in 1927. Carter also built a small putting course in the same romantic idiom of the surrounding grounds. Including winding paths, rock formations, and decorative statues of elves, gnomes, and fairies, this miniature golf course, the first of its kind, was patented in 1928 under the name Tom Thumb Golf Courses. Carter went on to sell six thousand miniature golf course kits before the company was sold in 1930.

The boom years of the Roaring Twenties were the last hurrah for the railroad mountain resort. The crash of 1929 rocked the world of tourism like a cataclysmic meteor, eradicating long-established patterns and paving the way for the rise of subdominant forms that were nonetheless better adapted to the new conditions. Like lumbering dinosaurs from an outdated age, many marginal mountain resorts faced extinction, along with their railroad lifelines. Scurrying in the shadow of the rail-based mass transportation, the private automobile and the public road rose to inherit the tourist landscape.

Carter's Fairyland Inn went bankrupt in 1931 and was sold to outside investors. Undaunted, the Carters turned to their ten-acre rock city, which Frieda had been grooming as her own personal rock garden since the late 1920s. Frieda devised the Enchanted Trail, which meandered through the rock formations past indigenous flowers and ceramic gnomes imported from Germany, culminating at a cliff-top prominence known as—what else—Lovers Leap, after two star-crossed Native American lovers forced over the edge by the stubborn chief of the Cherokee. Along the trail were such features as the Mushroom Rock, the Stone Witch, the obligatory Fat Man's Squeeze, Swing-Along Bridge, statuary depicting children's fairy tales, and a crevice stuffed with moonshining elves, a southern favorite. Rooted to the picturesque, with more than a hint of contrived Vaudeville, Rock City Gardens was classic road sideshow. The gates opened in 1932 to a growing, auto-bound clientele, and it has been an institution ever since.

NATURE AS RECREATION AND DIVERSION

The automobile changed the fundamental geography of nature tourism. Railroad attractions were site-specific, and tourist landscapes were assembled at a scale determined by distances that could be comfortably walked. By contrast the far-ranging flexibility of the automobile allowed entire regions to be assembled into tourist landscapes. Beginning in the late nineteenth century, this was done under the aegis of the National Park Service. Like the wilderness painters trying to capture the work of the sublime hand of God before it was replaced by human artifice, the federal government stepped in to preserve and protect huge tracts of remnant wilderness just as the last ridges were being scalped of their trees and disemboweled of their mineral riches.

Although conservation was a motivating factor, tourism and recreation were the overriding economic and political rationale for establishing national parks.[29] Nature tourism had experienced a decided shift away from the social events of the resort hotel, toward outdoor recreation that included hiking, boating, camping, and auto sightseeing. Established in 1916, the National Park Service emerged in the age of the tin-can tourist and the auto gypsy, when camping along the side of the road was a national pastime.[30] The National Park Service rounded up the most spectacular scenic wonders and packaged them in a way that was compatible with the auto tourist.

Shenandoah National Park in Virginia and Great Smoky Mountains National Park along the Tennessee/North Carolina border were projects that also fulfilled the desire to establish parks closer to population centers in the eastern United States. Right from the start it was anticipated that park visitors would arrive by auto. There was little choice. Although settled by mountain farmers and largely owned by lumber companies, both Shenandoah's Blue Ridge and much of the Great Smokies were remote islands of isolation that stood apart from the established passenger rail network.

Built along the crest of the Blue Ridge, Skyline Drive was the spine of Shenandoah National Park. South of Shenandoah the road continued as the Blue Ridge Parkway all the way to Great Smoky Mountains National Park, where it terminated at that park's main cross-mountain access road, the Newfound Gap Highway. This five-hundred-mile national parkway system was built as a recreational highway between 1936 and 1939.[31]

The parkways represented a radical departure from the utilitarian interstate highways the federal government also funded. Merging the picturesque landscape aesthetic with the machine-age technology of the automobile, the parkway was an extension of Frederick Law Olmsted's view of park space presented as a series of vistas.[32] It was the sinuous path through the garden employed at Rock City, Chimney Rock, Mammoth Cave, and other natural attractions, only at a scale appropriate for a leisurely drive, rather than a stroll. In the place of path-side benches and gazebos set for strollers to contemplate a particular view, the parkway had wayside pull-offs placed at scenic overlooks, and it emphasized rustic design elements such as stone facing and wooden guard rails. The parkways were enveloped within the restricted natural environment of a strip park so roadside access and vistas could be controlled and manipulated for maximum effect. The machine had moved beyond conquering the garden, it was now shaping it in its own image.

The recreational drives that meandered through the parks, with their rustic designs, controlled access, and scenic vistas, were in stark contrast to the commercial highways that led into the parks. Here utilitarianism, as defined by entrepreneurial capitalism, reigned supreme. The growing number of tourists that drove to

FIG. 16.5. The duality of American tourist landscapes is represented by the commercial utilitarianism of Gatlinburg's Parkway (top), in contrast to the restricted natural vistas of Skyline Drive (middle) and the Blue Ridge Parkway (bottom), which had essentially adapted the picturesque path to the landscape requirements of the automobile. Gatlinburg photo by the author; postcard views from the author's collection.

the parks required gas, food, and lodging, especially after World War II, when attendance doubled and more visitors were opting to stay in motels rather than camps. With severe restrictions on roadside development within the parks, commercial activities were pushed to the next nearest sites. This caused an explosion in the growth of gateway tourist towns, chief among them being Gatlinburg, Tennessee (figs. 16.5).

Overlooked by the passenger railroads, which scorned the nearly insurmountable grades of the Great Smoky Mountains Newfound Gap, pre-auto Gatlinburg was a "tiny mountain settlement unknown to all but a few lumbermen and their families, engineers, and lovers of wilderness camping."[33] By 1942 Gatlinburg had four hotels. A decade later the approach to Gatlinburg would be described as "a barrage of roadside signs, tourists' cabins, and gift shops that crowd on the heels of one another."[34] Gatlinburg, and its poorer cousin on the other side of the mountain, Cherokee, North Carolina, would emerge as the consummate examples of linearly arranged, auto-oriented gateway tourist towns. With the addition of their own contrived attractions—caged bears, sky rides and such—their diversions would eventually rival the park's own natural grandeur.

Buoyed by a booming economy, the postwar roadside-tourist world was a prosperous place for middle-class families on two-week vacations with station wagons filled with kids. The combination supported the vigorous proliferation of parasitic tourist attractions that attached themselves to larger destinations such as national parks or supplemented the revenue at smaller attractions such as commercial caves. These diversions were generally geared toward families with children whose appreciation, or tolerance, for long car rides and vistas was limited.

The aerial sky ride and the observation tower were common tourist diversions that merged the visual aspects of landscape appreciation with the family-fun thrill of an amusement park. Some sky rides carried tourists in enclosed gondolas, but most were displaced chairlifts more commonly associated with winter ski slopes. After the Gatlinburg Sky Lift opened to the top of Crockett Mountain in 1954, the sky ride spread throughout the scenic Upland South in epidemic proportions. Similar chairlifts soon appeared outside Mammoth Cave and Natural Bridge; in Maggie Valley, North Carolina, over the edge of the New River Gorge at Hawk's Nest, West Virginia; and up to the rim at Canyon Land Park, Alabama. Larger aerial tramways were built to the summit of Stone Mountain, Georgia, and up Mount Aetna, Tennessee, to supplement the revenues of Crystal Caverns.[35] Back in the foothills of the Great Smokies, additional sky rides were constructed at Ober Gatlinburg Ski Resort and Amusement Park, including a monster aerial tram that could accommodate 120 passengers in each car.

Predating the sky ride, the observation tower reflected the same goal, to generate revenue from scenic vistas that were otherwise free to be had. As a monadnock rising above the weathered northern Georgia piedmont, there is nothing surrounding Stone Mountain that even approaches its lofty heights. Yet as early as 1838 an observation tower was constructed at its summit, and tourists were compelled to mount the steps to add 30 more feet of viewing power to the 650 feet the mountain already provided. Although most have since fallen, the roadside observation tower was a staple of ridge-top one-stops that hawked gas, food, lodging, and diversion from the crest of every major transmountain highway from the 1920s through the 1940s. Their legacy is continued in the soaring, 342-foot Space Needle built over Gatlinburg in 1970 and the 216-foot Hot Springs Mountain Tower opened in 1983 as the latest in a series of towers that have stood on that summit since 1877.[36] In addition there are a number of more recent public observation decks set on such peaks as Mount Mitchell in North Carolina and Clingmans Dome in the Great Smokies.

As early as the 1950s, southern highland diversions started to lose their patina of regionalism, encouraging the reinterpretation of Mother Nature as just another competing tourist attraction. It began innocently enough in the 1960s, with the introduction of religious and Western theme parks. As an alternative to the handiwork of God presented by the Great Smokies, Gatlinburg's Christus Gardens opened in 1960 to display the handiwork, as diorama, of human interpretation of the Son of God. In the wake of the moribund mineral springs industry, Eureka Springs, Arkansas, reinvented itself utilizing the image of Jesus Christ after the Elna M. Smith Foundation's 1965 debut of the popular Passion Play outdoor drama. As a tourist destination Eureka Springs was born again to include a giant statue of Christ, a replica Holy Land, and similar religious diversions.[37]

The circular Tweetsie Railroad opened in the shadow of Blowing Rock, North Carolina, to preserve and present local railroad heritage, which centered on the extraction of timber from the Blue Ridge and had nothing to do with "Cowboys and Indians." Nonetheless a contrived frontier town right out of the Old West was brought to the trackside in 1961. The chairlift above Maggie Valley terminated at Ghost Town in the Sky, a Western theme park that also opened in 1961.[38] The Mammoth Cave equivalent was Guntown Mountain, where tourist tenderfoots rode a chairlift to see "realistic hangings" and other images of the storied West. Down in the Great Smokies, Frontier Land came to Cherokee, North Carolina, and Goldrush Junction opened in Pigeon Forge, Tennessee.

After World War II, electronic imagery, in the form of movies, television, and computers, came to be the dominant reflector and teacher of landscape interpretation. The proliferation of Western theme parks in the 1960s was an actualization of scenes and stories that were presented in the cinema as early as the 1930s and that continued as a staple of television programming up until the 1970s. Charting the rise and fall of the TV Western, Pigeon Forge's fading Goldrush Junction was used as the nub of another multi-media-inspired tourist diversion. In 1986 it was rebuilt as Dollywood, a romanticized version of the Great Smokies as interpreted by singer/ celebrity Dolly Parton. On the other side of the mountain, Cherokee's Frontier Land was bulldozed in the early 1980s to make way for Harrah's Cherokee Casino.

CONCLUSIONS

The appreciation of landscape paintings required long-term exposure to a static image, an activity mirrored by the tourist's voyeuristic appreciation of actual landscapes as seen from specific vantage points while in a position of repose encouraged through the provision of verandahs, gazebos, and benches. By contrast electronic imagery is dynamic, even schizophrenic, providing multiple and constant stimulus in quick succession. Throughout the twentieth century, technological image making progressed from the single-story cinema, through the increasing diversity of channel-driven television, to the unlimited, instantaneous, and infinitely diverse Internet. The evolution of tourist landscapes has mirrored this trajectory. Nature has gone from being perceived as the center-stage representation of the Almighty himself, to just one of many different tourist diversions. This does not mean to imply that people appreciate nature any less, just that there are many more options for vacationing tourists.

Since the ubiquitous expansion of limited-access interstate highways after 1956, reliable, comfortable, high-speed travel has become commonplace. With interstate travel the mountains are mere backdrops, sometimes far away, sometimes close at hand, but rarely an intrusion that would cause the speedometer to drop below sixty-five miles per hour.

The axis of movement and associated auto-oriented development has shifted from towns of the Dixie and Lee highways to the interchanges of Interstates 65, 75 and 81. These commercial nodes have eclipsed adjacent downtown retail districts in the provision of traveler services. Tourists vacationing at Mammoth Cave now have scant reason to travel U.S. 31E into Park City, Cave City, or Horse Cave, where old railroad hotels and early motels stand marginalized and abandoned. All their needs and desires for gas, food, lodging, and diversion are met by gateway interchanges out on I-65.

Despite the increase in contra-nature diversion, attendance at major scenic attractions such as national parks has increased. The diversity of recreation, however, has kept pace with the diversity of diversion out on the strip. The Generation-X youth culture touted by national advertisers has encouraged a nature experience defined by a myriad of high-intensity "extreme" (or maybe X-treme) sports. Whitewater rafting, snowboarding, mountain biking, off-slope skiing, free-style rock climbing, hang gliding, parasailing, and bungee jumping have all worked to reinterpret nature as an extreme diversion, or the recreational equivalent of a Mountain Dew commercial. Out on the strip the contra-nature diversions have followed suit with the spread of simulation theaters. For a nominal fee tourists can feel the simulated rush of shooting the rapids or plummet down a snowy cliff face strapped to a snowboard without the risk and discomfort required by a messy and sometimes uncooperative natural environment. In the end who is to say they are not, like the early pilgrims to the wilderness cathedral, experiencing the sublime?

NOTES

1. John A. Jakle, *The Tourist: Travel in Twentieth Century North America* (Lincoln, NE: U of Nebraska P, 1985), 65, and Stanford E. Demars, "Romanticism and American National Parks," *Journal of Cultural Geography* 11, no. 1 (1990): 21.
2. Edgar W. Spencer, *Natural Bridge and Natural Bridge Caverns* (Lexington, VA: Poorhouse Mountain Studios, 1985), 45.
3. E. P. Tompkins and J. Lee Davis, *The Natural Bridge and Its Historical Surroundings* (Natural Bridge, VA: Natural Bridge of Virginia, Inc., 1939).
4. Anna Marie Sobota, *The Story of the Natural Bridge of Virginia* (Natural Bridge, VA: Natural Bridge of Virginia, Inc., 1990).
5. Janet Valenza, "Taking the Waters at Texas Spas," *Journal of Cultural Geography* 11, no. 1 (1990): 57.
6. Betsy Blackmar and Elizabeth Cromley, "On the Verandah: Resorts of the Catskills," in *Victorian Resorts and Hotels: Essays from a Victorian Society Autumn Symposium,* ed. Richard Guy Wilson (Philadelphia, PA : Victorian Society in America, 1982), 51–52.
7. Ibid., 53.
8. William Olcott, *The Greenbrier Heritage: White Sulphur Springs, West Virginia* (White Sulphur Springs, WV: Arndt, Preston, Chapin, Lamb, and Keene, Inc., 1967), 19.
9. *Greenbrier County: 160th Anniversary, 1778–1938* (Greenbrier County, 1938).
10. Marie Cahill and W. J. Yenne, *The Pictorial Atlas of North America's National Parks* (Greenwich, CT: Brompton Books, 1993), 58–59.
11. Angela Miller, *The Empire of the Eye: Landscape Representation and American Cultural Politics, 1825–1875* (Ithaca, NY: Cornell UP), 50.
12. Barbara Novak, *Nature and Culture: American Landscape and Painting, 1825–1875* (New York: Oxford UP, 1980), 5–6.
13. Margaret M. Bridwell, *The Story of Mammoth Cave National Park, Kentucky,* 11th edition (Mammoth Cave, KY: Mammoth Cave National Park, 1971), 34.
14. Novak, *Nature and Culture,* 166–84.
15. Ibid., 176–84.
16. Kevin J. Patrick, "Transportation Corridor: The Development and Spatial Structure of Roadside Cultural Landscapes," *Small Town* 22 (January–February 1992): 4–11.
17. Ibid. Most likely this is the line that originates in Etowah, McMinn County, Tennessee.
18. Works Projects Administration, *Arkansas: A Guide to the State* (New York: Hastings House, 1941), 153.

19. Tompkins and Davis, *The Natural Bridge,* 48.

20. Martha Carver, "Dixie Highway Bus Tour Two: Moonpies, Monuments and Motels," in Martha Carver, Jeffrey L. Durbin, Claudette Stager, et al., *Drivin' the Dixie: Automobile Tourism in the South*" ([Nashville, TN]: Society for Commercial Archeology, 1998), section E, p. 54.

21. Howard Lawrence Preston, *Dirt Roads to Dixie: Accessibility and Modernization in the South, 1885–1935* (Knoxville: U of Tennessee P, 1991), 3.

22. Bridwell, *The Story of Mammoth Cave,* 39.

23. Patrick, "Transportation Corridor," 8.

24. Preston, *Dirt Roads to Dixie,* 104–5.

25. Tim Hollis, *Dixie before Disney: 100 Years of Roadside Fun* (Jackson, MS: UP of Mississippi, 1999), 140.

26. W.M. Cline Company postcard, Chattanooga, circa 1935, author's collection.

27. Item from Asheville Post Card Company, Asheville, NC, circa 1935, author's collection.

28. Carver, "Moonpies, Monuments and Motels," E-60–63.

29. Jakle, "The Tourist," 68.

30. Warren James Belasco, *Americans on the Road: From Autocamp to Motel, 1910–1945* (Cambridge: MIT P, 1979), 7–8.

31. *America's Highways, 1776–1976: A History of the Federal-Aid Program* (Washington, DC: U.S. Department of Transportation, Federal Highway Administration, 1976), 139.

32. Phil Patton, *Open Road: A Celebration of the American Highway* (New York: Simon and Schuster, 1986), 68.

33. Laura Thornborough, *The Great Smoky Mountains* (1937; reprint, Knoxville, TN: U of Tennessee P, 1956), 49.

34. Ibid., 49.

35. Today Mt. Aetna is better known as Raccoon Mountain.

36. Hollis, *Dixie before Disney,* 71–72.

37. Ibid., 72–73.

38. Ibid., 162–69.

CHAPTER 17

PILGRIMAGE ROUTE TO PARADISE: THE SACRED AND PROFANE ALONG THE DIXIE HIGHWAY

ROBERT M. CRAIG

Throughout the centuries people have traveled to places nearby or faraway in order to visit friends or relatives, to shop, to conduct business, or to relax. Whatever the destination, the wayfarers had stories to tell about the trip itself, the people they met along the way, the sights they saw, and the roads they traveled. Robert M. Craig compares characters in one of our best-known narratives, Chaucer's *Canterbury Tales,* with early-twentieth-century tourists motoring along the Dixie Highway and with our present-day efforts to recapture the spirit of the Dixie. The author's tongue-in-cheek comparison between the routes to "paradise"—Canterbury and Miami—in medieval times and the twentieth century, respectively, gives a glimpse not only of the destinations, but also of adventures on the way. Whether it was finding places to sleep and eat, purchasing reliquaries, or visiting Rock City or the Fountain of Youth, the pilgrims' experiences along the journey were as interesting as reaching the endpoint of the trip.

Longue er April with his showres soote
The snows of Janus cover trunk and roote,
And bathed frigid earth in white licour,
Of which vertu is hidden all the flowr;

. . . Thanne longen folk to goon on pilgrimages,
And palmeres for to seeken straunge strondes
To ferne halwes, couthe in sondry londes;
From Soulte Sainte M'rie at very shires ende
On Dixie Highway to Miami wende,
Unholy sunful paradise to seeke
That Fisher builden under winter heete.

After Geoffrey Chaucer's
Canterbury Tales

From the 1920s to well after World War II, the Dixie Highway was the U.S. pilgrimage route to Miami, a secular road to Santiago da Compostela. Paradise-bound sun worshipers followed the Eastern and Western divisions of the Dixie Highway, much as twelfth-century pilgrims had set out from Vézelay or Paris or Arles, heading south on pilgrimage. Dixie Highway travelers crossed the Appalachians, as their kindred pilgrims had crossed the Pyrénées, and traveled not to the shrine of St. James but to Miami Beach, the shrine of "San Carlo" Fisher.

Both pilgrimage routes passed through miles of rural farm country, the earlier, medieval landscape dotted with timber-roofed, stone tithe barns, those massive secular cathedrals wherein was stored the abbot's wealth from the feudal farm land. The other pilgrimage route, to Dixie, was lined with farmsteads whose barns invited pilgrims to "Chew Mail Pouch" or to "See Rock City," the latter, a garden of rocks, not a city, and a shrine to advertising overkill.[1] One might easily believe a holiday trip south through the Bible Belt would, or should, be sprinkled with religious experiences. U.S. 41 brought the Dixie Highway through Tennessee where Wonder Cave, the great secular rock of ages in the mountains near Monteagle, invited travelers to "See Priest at the Altar." Still further south along the Dixie Highway were makeshift billboards advertising southern barbecue, where golden calves and pigs met the same fate of martyrdom as St. Vincent, and no pilgrim could miss the omnipresent white cruciform roadside shrines to Burma Shave:

Within this vale
of toil and sin
your head grows bald
but not your chin
Burma Shave

Twentieth-century pilgrims along the Dixie Highway carried oil company routing booklets, Dixie Highway strip maps, and AAA tourist guides to hostels, collectively constituting a modern-day *Liber Sancti Jacobi* or Codex *Calixtinus*.[2] The pilgrimage to paradise had little changed since the twelfth century.

It would be sacrilegious to suggest that the medieval pilgrimage road of the Romanesque era was the birthplace of the tourist trap. The Romanesque route to Santiago was punctuated with pilgrimage churches, often a mere day's journey apart, but can it be said that Moissac and Souillac were merely roadside markets, institutional traders in souls and purveyors in absolution for a price? Martin Luther thought so. At St. Sernin in Toulouse and elsewhere along the road to Santiago, priests offered indulgences, clerics bartered for the remission of sins, and pilgrims became history's first a-Faustian gift-shop scavengers (that is, they would buy anything to save their souls). The exchange of money for goods in the Middle Ages reaped more than a pecan log or one of life's other indulgences at a Stuckeys outlet: the rewards for viewing relics and the credit accrued from purchase of indulgences in medieval Europe would sustain the pilgrim and promised to secure the believer's entry into paradise. Seeking Santiago was symbolic of the whole earthly sojourn: the road to paradise was, after all, the ultimate of life's journeys.

Therefore medieval travelers stopped at every roadside attraction to view the contents of side shows and reliquaries, and they collected and traded pilgrimage artifacts of extraordinary variety. The pilgrimage church became a veritable marketplace on festival day, offering relics and holy trinkets (from bones to thorns to cloth), each sanctified by association regardless of the fact that the authenticity of some relics stretched the credulity of even the truest believer. Chaucer describes his "Pardoner," on the road to Canterbury, as a peddler in relics, those passports to paradise, which he had hidden away in his bag or pillowcase (male):

> For in his male he hadde a pilwe-beer
> Which that he saide was Oure Lady veil;
> He saide he hadde a gobet of the sail
> That Sainte Peter hadde whan that he wente
> Upon the see, til Jesu Crist him hente.
> He hadde a crois of laton, ful of stones,
> And in a glas he hadde pigges bones.[3]

As the great secular pilgrimage route south, the Dixie Highway featured museum reliquaries of the most imaginative local collections and attractions. These, as promised the medieval depositories of relics, paved the way to the eventual entry of the pilgrim into paradise that Carl Fisher was developing just beyond the Everglades. Like the purchaser of medieval indulgences hoping to erase the sins of time with a guarantee of eternal life, thousands of Dixie pilgrims stopped along the Eastern Division of the Dixie Highway in St. Augustine, Florida, to drink from the Fountain of Youth; this was in spite of the lack of even a shred of evidence that the waters had restorative powers. The Fountain of Youth was immensely popular. Winter refugees, who had driven for hours or even days, sensed, as they finally crossed the Florida state line, that they had entered paradise and that already they felt younger. Ironically another Fountain of Youth in St. Petersburg, Florida, was nothing more than a municipal fountain at an artesian well where the faithful drew water daily and reported that they did indeed feel better. Perhaps there was good reason: when the Pinellas County Health Department tested the water in 1971 officials found a high percentage of the trace element lithium.[4]

Santiago's *camino real* had nothing over the Dixie Highway as an *allée des trésors noirs* ("alley of black treasures"). Most Dixie Highway roadside attractions in the South were family founded and mom and pop managed and thus promised more wholesome experiences than could be found at "wicked Las Vegas."[5] Ironically, a small strip in the area of Kimball, Tennessee, has been named "Little Las Vegas" as the nearby fireworks plant at South Pittsburg and the Atomic Fireworks Space Age Gift Shop and the Tennessee-Alabama Fireworks and Gift Shop concentrate so much colorful lighting and neon that the night sky is ablaze. One suspects that one careless match could create a convincing Armageddon. One must also confess, however, that some Dixie pilgrimage route attractions were as gruesome to the faint of heart as the reliquary-encased index finger of an apostle. Dixie travelers were always assured of thrills, surprises, and attractions beyond belief, a marketing pledge still alive at Ripley's Believe It or Not in St. Augustine and elsewhere. On the road to Santiago, there were souvenirs, preserved saints' bones, and saintly organs that would rival a modern day "House of Horrors," and enough pieces of the True Cross to construct a cathedral roof.

> But with thise relikes whan that he foond
> A poore person dwelling upon lond,
> Upon a day he gat him more moneye
> Than that the person gat in monthes twaye;
> And thus with feined flaterye and japes
> He made the person and the peple his apes.[6]

Moreover, the pilgrimage route to the temples of the Florida sun included at least one shrine to martyrdom along the Eastern Division of the Dixie Highway, the Tragedy in U.S. History Museum in St. Augustine, where pilgrims were enticed to "See Jayne Mansfield's Death Car" and "learn the truth." Deaths from automobile accidents on U.S. roads were occurring at increased frequency. Eventually southerners began to mark locations of deaths from traffic accidents with floral shrines along the highway, a distant echo of the tombs lining the Appian Way or of Romanesque martyrium. Harrison Mayes of Middlesboro, Kentucky, was not land-marking tragedy and death so much as warning of the Last Judgment when he erected hundreds of concrete crosses, primarily throughout the South. Messages included "GET RIGHT WITH GOD," "JESUS IS COMING SOON," and "NOTHING BUT THE BLOOD."

Just off the Dixie Highway in central Florida, an isolated tower, in the tradition of monumental Romanesque church towers, points heavenward as it overlooks the low, rolling topography of the orange groves, the holy vineyards of paradise. Recalling a medieval campanile, Bok Tower, built in 1927 and sometimes known as Singing Tower, is a landmark of Art Deco design. The surrounding grounds were formed into a garden oasis by the Olmsted Brothers, the successor firm composed of the sons of the renowned landscape architect Frederic Law Olmsted Sr. The brass repoussé north door of the tower depicts the six days of Creation, and the marble grill of the west window presents traceries of a man creating a garden. The Singing Tower and Sanctuary have become a memorial to the former editor of *Ladies Home Journal* and founder of the sanctuary, Edward Bok, who, like a medieval church-building abbot, completed his work and then died, in 1930, within sight of his memorial; he is buried at the base of the tower. As the carillon plays at dusk, Bok Sanctuary becomes its own quiet paradise; as one historian described it, "the Sanctuary truly seems to come alive and sing its creator's gospel of beauty to whoever will take the time to listen."[7]

> I, master Gonzalo, from the village of Berceo,
> While on pilgrimage chanced upon a meadow,
> Green, untrodden, full of flowers.
> Just a resting-place for an exhausted traveler.[8]

Of all the wayside shrines along the Dixie Highway pilgrimage route, Bok Tower may be the holiest in its capacity to instill quiet and referential communion and self-reflection. Once a sanctuary for the Native Americans of central Florida, Bok Tower traces the idea of a sanctuary primarily for human meditation to "the great Aryan Sanctuaries of India," presenting itself in a tradition "continuous since primitive man first sought peace within." Edward Bok's sanctuary is one of the most serene pilgrimage sites of the South (fig. 17.1).

If the Dixie Highway brought winter "snow birds" south to experience paradise, then the best attraction was nature herself. With the finest landscape architects in the country, Bok created his own Edenic sanctuary, but there were other Gardens of Eden in Florida at the end of the pilgrimage road. Some were called Eden, others were jungle gardens or parks, and still others were natural springs or preserved and developed lakes of noted natural beauty. These slices of nature's great "unspoilt paradise" were found along both the Dixie Highway's Eastern Division in Florida (U.S. 1) and its Western Division (U.S. 17, U.S. 27, U.S. 41): at the Garden of Eden in Palm Beach; at Cypress Gardens, founded in 1936 near Winter Haven; and at Silver Springs near Ocala.[9]

FIG. 17.1. Bok Singing Tower, Mountain Lake Sanctuary, Lake Wales, Florida, built 1927. Milton B. Medary, architect; Lee Lawrie, sculptor; Frederick Law Olmsted, landscape architect. Photo by the author.

This last-named shrine to Edenic paradise, Silver Springs, had attracted hundreds of tourists long before the Dixie Highway traversed this part of central Florida. Silver Springs was a group of natural springs that travelers originally accessed by means of a steamboat chugging up the Ocklawaha River.[10] The whole probably resembled a scene out of the 1951 film *The African Queen.*[11] The experience was so full of mystery and steamy discomfort, as the "clammy, cramped, and bug-ridden" boat propelled itself through the alligator-infested waters of a jungle, that pilgrims may well have believed it necessary first to pass through purgatory before entering paradise. But as the natural spring water grew clear and light blue, the seekers after the waters experienced what a particularly introspective pilgrim described as "one of the most dramatic transitions from darkness into light that a traveler can make anywhere on the continent."[12] Margot Ammidown has noted that "many of the written descriptions of the early trips to the springs seemed to equate the journey with a spiritual transition to the afterlife, or to refer to the time-honored notion of the river as a metaphor for a spiritual journey to the source, which, with the advent of tourism, became a regular mini-pageant acted out on the Ocklawaha."[13]

The journey through paradise also brought expression along the Dixie Highway to a Florida still to be experienced as a tropical jungle, an Eden before civilization (that is, before development). Long before bevies of flamingos greeted visitors at the entries to theme parks, Florida preserved or developed jungle gardens, with many such attractions dating from the mid-1920s and 1930s and most located in southern Florida. Here, too, pseudo-religious references and ecclesiastical metaphors abound. On the west coast the Sarasota Jungle Gardens offered pilgrims "The Prayer of the Woods" entreating visitors to "harm me not," because "I am the wood of your cradle and the shell of your coffin." McKee Jungle Garden in Vero Beach was atypical in not being a reclaimed and tamed jungle but the new design of the landscape architect William Lyman Phillips. It included a jungle trail to a "cathedral of Royal Palms."[14] Miami's Serpentarium dates from 1947; where would Eden be without a serpent?

Gatorland Zoo in Kissimmee, built in 1950, was merely one of a long line of reptile farms stretching the full length of the peninsula, from Alligator Farm in St. Augustine to Alligator Joe's in Miami and Palm Beach.[15] Joe Campbell and Henry "Sonny" Coppinger Jr. captured alligators by hand, put them to sleep by rubbing the reptiles' bellies, and introduced the roadside attraction of "wrestling alligators," creating what one historian has called "South Florida's version of Saint George wrestling the dragon."[16] Mythological and pseudo-religious experiences were not lacking along the pilgrimage road to San Carlo alle Tamiami.

Safari tours of the Everglades, displays of Seminole culture, gator wrestling, and jungle parks seemed to proliferate the further south in Florida that the Dixie Highway traveler progressed. There was little that was threatening about this untamed wilderness. Indeed a subliminal message was being conveyed to Dixie Highway pilgrims concerning the transformation underway in paradise. Appearing on neither billboard nor venue marquis, but perceived by travelers nevertheless, was the demonstration of humanity's dominion over the earth: at the end of the pilgrimage route, nature was being tamed, and eventually developers would civilize Florida. The final destination of Dixie Highway pilgrims was Miami, where they could bear witness to this transformation at miraculous speed, from mangrove swamp and jungle wilderness to hedonistic paradise. The curiosities along the route might range from Tom Gaskins's collection of cypress knees, begun in 1934, to Cherokee or Seminole artifacts and crafts strewn along the road like a linear garage sale, to airboat rides across the tall grass and water of the Everglades, to Frog City along the Tamiami Trail. The earthly pleasures promised by promotional advertisements along the way included bathing beauties, orange juice that ran like water out of the taps, and the great elixir, eternal sunshine, as vital as the Chinese Ch'i, or breath of life.

The Dixie Highway led south to this "land of sunshine," and billboards every hundred feet repeatedly promised the world's largest or the world's smallest or the most fantastic or the most curious artifact of one kind or another. Hungry pilgrims driving toward Miami approached the Old South Barbecue; indeed, for more than a hundred miles north of the gastronomic shrine, innumerable billboards spaced at regular intervals reminded travelers that the Old South Barbecue promised the "world's finest" pork from a spit. As one got closer to the hallowed grounds of swinish gluttony, one began to pass Old South Barbecue billboards at more frequent intervals, with their message now giving a countdown of the remaining miles to point zero, a preparation for peptic palpitations worthy of Cape Canaveral. And then the most understated wayside shack slipped by, and, for several miles further on, billboards started reappearing to alert the pilgrim in large letters that "YOU MISSED IT!" and "YOU'VE DRIVEN PAST," demanding that the traveler "GO BACK." In the tradition of Old South Barbecue advertising, South of the Border on I-95 and "topless" truck stops on I-75 continue today the roadside advertisement media blitz with the same intentions: to stop the south-bound motorist for a little indulgence before entering paradise.

Old South roadside barbecue eateries were often mere rustic sheds with little architecturally to recommend them. Fireworks shacks, makeshift gift shops, and less-than-rustic tourist-camp cabins were little better, looking much like housing for retired alligator wrestlers. The earliest way stations of Dixie were often undistinguished, ramshackle shanties, and if the southern back kitchen sometimes looked too much like an insalubrious shack, tourists sometimes traveled on to an atmosphere a bit more upscale, or at least an environment more conducive to ease of digestion as the pilgrim dined en route.

What does stand out nationally as designed (not merely slapped together) and peculiarly "of the road" are roadside diners and streamlined gas stations, where, since the 1930s, both passengers and automobile have "tanked up." First constructed in New England and the Middle Atlantic states, such buildings began to appear southward and westward along U.S. 1, Route 66, the Dixie Highway, and other national roads, creating an increasingly universal aesthetic of the thirties, soon to constitute the quintessential image of roadside architecture. The Streamlined Moderne was, indeed, an aesthetic *of* the road. Its architectural features derived from curved corners, lines of speed, and the metal shrouding of streamlined locomotives, airplanes, coupes, and roadsters of the period. The Dixie Highway is contemporary with the rise of the car culture nationally. Just as we associate late Art Deco and Moderne aesthetics with transportation, we perceive the early motel, the roadside diner, standardized Texaco service stations, and the lines of speed informing each, with the increased efficiency, reduction of time and space, and modernity of new roads such as the Dixie Highway. Those roads, of which the Dixie Highway was the major route south, were a significant part of the national experience from the 1920s on.

Medieval pilgrims *walked* to Santiago, and the rhythm of aisle bays, ambulatory, and radiating chapels encircling a reliquary in a Romanesque pilgrimage church between Vézelay and the Shrine of St. James reflects the gait of the *perambulating* pilgrim. The best pilgrimage travelers could do (since human feet set the pace) was to increase speed on horseback. Much later, people could get from point A to point B with greater speed in railroads and automobiles whose engines were measured by horsepower. When the Dixie Highway Association first met in 1915, it was clear to those such as Carl Fisher, anxious to get travelers all the way to Miami, that if roads would improve, the pace of the pilgrimage could be increased substantially from that of Chaucer's or Master Gonzalo's day.

The Dixie Highway system brought significant progress in the building of better roads during the period between the world wars. In 1916 the association published its first official Dixie Highway map, and national estimates indicated that road improvements enabled travelers to cover 125 miles in a day's journey. Chaucer's entourage could barely make that distance in a week. The 1916 figure of 125 daily miles was increased to 170 miles per day in 1920, and 200 miles per day by 1925, the tenth anniversary of the Dixie Highway. Motorists could cover 300 miles per day by 1931, and 400 miles per day in 1936, the year Carl Fisher died.[17]

The Dixie Highway began to signify speed and ease of travel. Lining the highway were motels and diners and gas stations that were increasingly accented with neon and designed in the new aesthetic of streamlining. Such buildings began to shape an aesthetic of the road, an architecture of speed and modern efficiency that itself has become an historic image of the 1930s and 1940s. It should be observed, therefore, that beyond the phenomenon of roadside attractions, much roadside architecture of the period styled itself in the streamlined phase of Art Deco–Moderne that was distinctively "of the road." In so doing, Route 66 and Dixie Highway roadside architecture helped to define the characteristic aesthetic of U.S. car culture. In this respect the Dixie Highway was a sign of the times, to borrow a phrase from Thomas Carlisle.

During this era, from the 1920s to the 1950s, the iconography of Americana changed. For Grant Wood, Middle America was still defined in nineteenth-century terms as a farmer and his wife posing in front of a rural vernacular Gothic farmhouse, an image of isolation that by 1930 was already breaking down as a result of the network of national roads crisscrossing the country. Wood's 1930 painting "American Gothic" became an icon when it later appeared on corn flakes boxes even as George Lucas's 1973 film *American Graffiti* was becoming an equivalent iconic summation of life for mid-twentieth-century U.S. youth and the car culture. Grant Wood's farmer and his wife were sturdy stock and static, conservatively spartan, terse, and laconic, reminiscent of Calvin Coolidge. And they were Republicans. The teenagers of *American Graffiti,* on the other hand found freedom, on the edge of license, in the automobile. They viewed the car as a socializing instrument offering independence from the mores of polite society. They were full of energy, loose, liberated, and liberal. And they were Democrats.

For twentieth-century youth, the signs of the new times, the "good times," were the hot rod and the dragstrip racetrack, the rituals of lover's lane parking, "pick-ups" [meaning either trucks, or what the British call cheap tarts], pit stops at neon-lighted drive-ins, and the flashy styling of commercial strips. These reflections and embodiments of the car culture witnessed the coming of age of the teenage driver and became emblems for generations of youthful free spirits, intact even after the automobile killed James Dean on the highway. The Dixie Highway was a road well traveled: Grant Wood's farmer could have sojourned to (or retired in) the Miami Beach of 1930; the youthful rediscovery of "South Beach Art Deco District" in Miami Beach fifty years later reflects both the birth and revival of U.S. car culture and Streamlined Moderne design that had first appeared as roadside architecture.

If the phrase roadside architecture, rather than roadside building, is to be employed, the reader may conclude the buildings to be discussed are not tumble-down rickety way stations but structures of noteworthy aesthetic design quality. Bok Tower is architecture; a fireworks stand or the generic southern barbecue eatery may not be. To be sure, the archaeology of U.S. roads uncovers artifacts of culture that are more egalitarian than elite, and more everyday vernacular than high-style Sunday best. Snobs argue the ordinary structures of "pop cultural" studies—gas stations, diners, motels, and the like—are not architecture at all, but mere building, forgetting that John Ruskin never drove a car.

We can credit (or blame) John Ruskin for cementing such distinctions in the collective cultural mind-set. Ruskin's admonition that "when we build, let us think that we build forever"[18] would have fallen on deaf ears in the twentieth-century culture of planned obsolescence, sound-bite marketing, and fickle fashions, all of which contributed to the spiraling of aesthetic tastes deeper into the abyss of consumerism and the marketplace. Ruskin would be appalled with Las Vegas and mortified by the generic suburban commercial strip.

Yet Ruskin, too, saw buildings as autobiographies of culture and societies, and he argued that the book of art was more reliable than books of deeds or books of words.[19] Nevertheless, despite Robert Venturi, I doubt a modern-day Ruskin would ever step foot in an A&P parking lot or stop for high tea at an A&W Root Beer Stand. Regardless of his socialist protestations, Ruskin was a snob. And if he were alive today, he would offer the politically incorrect view that a high degree of discrimination should color our surveys of roadside buildings. He would utter the blasphemy that there may be icons of the U.S. highway that are better, not just older, than others, that there are artifacts of the roadside of superior aesthetic merit, maybe even more meaningful as embodiments of our car culture, than others. This is not an inclusivist's view, and in his "thrice denial" Ruskin presents himself as the doubting Thomas of Society for Commercial Archeology (SCA) devotion.

There are some readers who would claim to be as aesthetically moved by a newly restored Cord 810 parked out in front of a refurbished Worcester Lunch Car Company diner as was Ruskin by the Doge's Palace in Venice or Giotto's Campanile in Florence. But travelers in Ruskin's day took along *The Stones of Venice* as a guide, and *The Seven Lamps* as a bible, not the SCA's Dixie Highway tour guide mapping out bedspreads, bypasses, MoonPies, monuments, and motels. After several years of madness, Ruskin died in 1900, just as the twentieth century loomed; perhaps he knew what was in store, aesthetically, further down the road.

Ruskin had nothing on the SCA for approaching architecture with a religious fervor, and if his pronouncements were considered by detractors to be the ravings of an aesthetic zealot, late-twentieth-century devotees of roadside architecture embrace "the roadside cause" with comparable veneration. We members of the SCA gather periodically for roadside conferences, which outsiders, who do not understand our particular devotion to the ordinary and ugly, refer to as the weekend that we get our annual car-culture fix.

I prefer to think of our devotional gatherings as a roadster revival meeting, a kind of campground religious experience. Here we seasonally seek absolution for societal sins of demolition, by calling with raised arms and bended knee for the salvation of threatened but revered "ordinary" buildings. At the annual SCA pilgrimage missions, we, too, worship relics and purchase Deco indulgences: miniature Miami Beach hotels that we can wear as ceramic broaches to ward off the evil spirits of the unenlightened, or SCA transfer-adorned coffee mugs to line up atop our office desks like votive candles to serve as reliquaries for our streamlined Parker pens. Scholarly papers are delivered by itinerant evangelists of the highway who here, in *this* congregation, preach the gospel of speed only to move on, some other holy day, to the Southeastern American Society for the Eighteenth-Century Studies, where periwigged savants, with equal aplomb, discuss rococo cartouches and other topics far removed from Hupmobiles or Dixie Highway motels. Whatever one's denominational preferences, aesthetics can be a theology: free, liberal, orthodox, or evangelical. On a diner weekend tour or an SCA conference field trip, we are pilgrims in the spirit of the Wife of Bath, of Chaucer's Pardoner, or of an early motorist along the Dixie Highway. The SCA has become the Streamlined Church of the Automobile.[20]

Therefore, in the spirit of the southern revival, this essay concludes with a testimony of faith. If the reader is convinced that Art Deco and the Moderne have gone the way of old time religion, that the values held dear, as embodied in 1920s–1940s diners, gas stations, and cinemas, are now somehow lost in the catacombs of a past history, that the nonconverted in state departments of transportation, who are protagonists of street widening and new development, have already razed most of the holy shrines of the Dixie Highway without even a genuflection, let me assure you that THE MODERNE HAS BEEN BORN AGAIN. It lives on, along the byways of Dixie, presenting itself as the neon "light of the world" as fervently as a way station for born-again Christians. Yet, even after new roads bypass sections of the Dixie Highway to isolate and abandon mom-and pop motels, that after Dixie Café, Dixie Auto, Dixie Camp, and Dixie Motel have lost their meaning, that during the last ten years of the twentieth century, Dixie diner Deco has been *born again* along southern roads.[21] Hallelujah! A resurrected Streamlined Moderne has become the Lazarus of the highway, razed from the dearth of diner deliverance and development death knells, to rise again as the neo-Moderne.

FIG. 17.2. Hot Rod Café, Woodstock, Georgia, 1995. Photo by the author.

The "Shrine of the Neo-Moderne" is in Woodstock, Georgia, a few miles off the Dixie Highway. There you may find, at car rallies on alternate Saturdays, a congregation driving roadsters and coupes, gathered in homage to the Hot Rod Café. This Dixie Deco diner dates from 1995, and it is among the most recent holy shrines of the Streamlined Church of the Automobile. Should the reader choose to make the pilgrimage, he or she will first approach the parking lot, the Moderne *parvis,* or forecourt, for pilgrims to gather before entering the shrine. The great portal, recalling splayed and tympanum-enriched Romanesque church entries, is here a chromium-arched portal echoing the profile of a juke box.

The juke box is itself a reliquary of audio chants by modern-day Gregorians whose evensong inside is either induced by overenthusiastic libation caused by too many trips to the altar rail, or by mystic séance, but from whatever cause, evensong can sometimes promote among the flock an urge to dance and cavort while speaking in tongues. With this in mind, the Hot Rod Café frescoes its exterior walls with icons of the new religion, prominent of which is the saint of jazz, Betty Boop. Among the stations of the cross is the signpost of the road to paradise, a Route 66 highway sign. The various instruments of the saints are displayed, including the guitar and microphone stands, and the *a secco* artist has included images of communicants enraptured in liturgical rites of passage, some sectarians line dancing, others jitter bugging. These images on glass, in the spirit of Ruskin's concept of the church as a Bible in stone, contribute to the creation of a neo-Moderne Bible in aluminum, chrome, and stainless steel.

Inside the Hot Rod Café are images of the saints: primary among which are typically the not-so-holy trinity of Marlon Brando, James Dean (fig. 17.3), and the Mary Magdalene of the Hot Rod Shrine, Marilyn Monroe. The King of Kings, of course, is Elvis Presley, while it should be noted that lesser and later saints such as Jack Nicholson appear on the walls, for instance, of Roy's Grill, a 1949 diner located on the Western Division of the Dixie Highway in Rossville, Georgia, and a site of the 1998 pilgrimage. At the Hot Rod Café, the bishop's chair, or *cathedra,* is an upholstered front seat of a horseless chariot, whose armrest is a chrome-bejeweled fender and fin from an unidentified 1950s automobile; the presence of

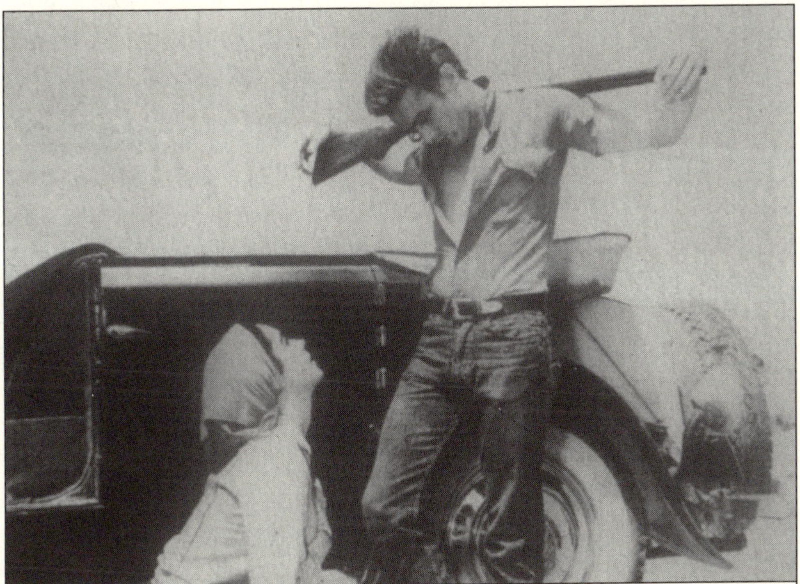

FIG. 17.3. Film still from *Giant:* James Dean in "crucifixion stance." *Giant* © Giant Productions. Licenced by Warner Bros. Entertainment, Inc. All rights reserved.

FIG. 17.4. The Cathedra, Hot Rod Café, Woodstock, Georgia, 1995. Photo by the author.

a cathedra, of course, indicates to the informed that this is the mother church of the diocese (fig. 17.4). And behold the holy of holies itself: the Hot Rod, an object of worship for the now aged generation of *American Graffiti,* a chariot of fire parked just off the ambulatory where ritualistic dances to *karaoke* are performed by those who knew of Ron Howard before he became a director.

The Hot Rod Café has taken great care to furnish its interior with all the vestments and sacred vessels of the occasion. Waitresses are frocked, as it were, in pleated skirts and bobby socks, and orders may not be taken without the proper habit. The bar stools are liturgically correct, and table linen is mandated to be gingham red-and-white check, the red presumably for the devil, the white for angels, set in balance as reminders of the unpredictable Last Judgment. The holy water is properly stored in metal coolers, the sanctioned ice chests for such libation. The curates of the countertop, ever vigilant to ensure propriety of office, have displayed only the authenticated straw dispensers, napkin holders, salts and peppers, and the like. The counter is a virtual altar rail adorned with soda-pop bottles in iconic reference to the innocence of youth (fig. 17.5).

If the Hot Rod Café may be considered the diocesan headquarters of the cult of the neo-Moderne, the pilgrimage routes are aligned with sister shrines embodying like features and iconic signs of the architecture of the automobile. Like disciples spreading the gospel from mission to mission throughout the South, the most prolific exemplars of the new aesthetic are Sonic, Rally's, and Checkers drive-ins. Found on and off the still-used portions of the Dixie Highway, these late-century

FIG. 17.5. The Altar Rail, Hot Rod Café, Woodstock, Georgia, 1995. Photo by the author.

roadside stopping places along the pilgrimage route are for travelers *en séjour* from work to home, or from Sault Ste. Marie to Miami. The drive-ins incorporate signs of the monastic order of the road by referencing the colors, patterns, and character of the Moderne revival. Fire-engine red is loud and shrill and forces pilgrims to notice. Black-and-white checkerboards symbolize the speed of the highway as race-track. At night neon light attracts even the most lethargic communicant, like photo-tactic insects, who is drawn off the Dixie Highway toward a vision of the Light of Life, the vibrant and undulating neon. Checkers Drive-Ins are mere kitchens with flanking porte cocheres, but we should note they are all appropriately cruciform in plan. Streams of pilgrims line driveways, like nave aisles, moving toward transept arms under whose vaults the exchange of indulgences takes place. The streamlined neo-Moderne drive-thru formula is as functional today as the Romanesque pilgrim-age church was in the twelfth century, and much as Cluny spawned Burgundian Cluniac off-shoots, so Rally's spawned Checkers, and countless cruciform Dixie Diner Deco drive-thrus now line the suburban highways of America.

Some adopt the language of stainless steel, portholes, and horizontal lines of speed. Rockin' Robbins produced two reincarnations of diner Deco in a line of diners whose name pays homage to the patron saint of *American Graffiti*, Ron Howard. These are called "Happy Days Diners" (fig. 17.6), a production line of Module Mobil, which has produced more than a thousand diners in various forms nation-wide since 1959. Lilburn, Georgia's Rockin' Robbin of 1988 is the first of their Happy Days Diners line, although we should note further that the name Rockin'

Robbin recalls Bobby Day's no. 2 hit tune from 1958. A second Rockin' Robbin was built in Snellville, Georgia, in 1992–93.

The glitzy, reflective metal of these neo-Moderne diners found more expressive character in the more haute-cuisine diners of the higher rent district of Buckhead in Atlanta. Landmark Diner of 1993 and more especially Buckhead Diner of 1987 are contemporary with the Rockin' Robbins diners but one quickly finds Buckhead Diner to be a Star Wars refectory for moneyed prelates rather than novice initiates who have sworn the oath of poverty. Buckhead radiates layered overhangs of mirrored reflection as brilliant as the chrome bumpers of mid-1950s DeSoto coupes.

FIG. 17.6. Rockin' Robins, a "Happy Days" diner, Indian Trail, Atlanta, Georgia, 1988. Photo by the author.

The distracting richness of Buckhead Diner was as likely to cause a reaction, as Cluniac art did for twelfth-century Cistercian conservatives. Cistercians sought a less worldly aesthetic, and a comparable and simpler approach to historic roadside diners of the 1990s found refuge in historic White Castles and Steak and Shakes as models stripped of the abhorrent Cluniac ornament. Virginal white skins displaced the bejeweled reliquaries of stainless steel and Cluniac chrome to encourage a less ostentatious roadside aesthetic in Johnny Rockets Hamburger chains (fig. 17.7). Art historians would conclude that Johnny must have been of Cistercian persuasion: his use of rounded corners, clean lines, and ribbon windows announce the borrowed Moderne but within the pure, almost abstract language of an early modern, a stripped-down simplicity that would be more acceptable to St. Bernard of Clairvaux. Johnny Rockets restrained interiors encourage a nostalgic recollection of Edward Hopper's *Night Hawks* and in so doing demonstrate that they are of the true faith. At this revival, counter service, short-order cooking, and all the reproductions available from suppliers of neo-Moderne accoutrements create a stage set for middle-aged baby boomers to recall with nostalgia the palpitations of their first date.

FIG. 17.7. . Johnny Rocket Hamburgers, Atlanta, Georgia, Sherrod Marshall, design concept circa 1987. Photo by the author.

A masterpiece of fifties revival can close the inventory of neo-Moderne landmarks along the Dixie. Its formal source is less the Streamlined Moderne of the 1930s than the Tastee Freezes and jet-aged Moderne of the 1950s. But the roadside fast-food emporium dates from 1991; it is Zestos, on Ponce de Leon Avenue in Atlanta, barely a decade old, but it joins the Shrines of the not so Immaculate Consumption that have been documented here.

Zestos is a serious piece of revival architecture. Diner expert Richard Gutman consulted with architect Dennis J. Dubey to create an authentic expression of roadside historicism; they took the owner's charge to maintain the spirit of the 1950s as a mandate and approached their task with a moral conviction worthy of Ruskin. Inside, architectural and decorative features were as carefully studied as if they were installing an academic period room in a museum. Zestos's use of cove ceilings, booths with music boxes, tile floors, stainless steel short-order cooking behind counters, and other features are rooted in the diner tradition. Zestos is the ultimate of populist egalitarianism. Its counters display ordinary checkerboard tile work whose brilliant yellow and red colors employ a palette derived from soft plastic squeezable mustard and ketchup dispensers. With as little faith as a mustard seed, Zestos canonizes a mere building as architecture: never was so ordinary an object of popular culture as a plastic tubular condiment container converted into a device of architectural ornament and craftsman detail.

In the 1920s the Dixie Highway was an Art Deco–era road to a hedonistic Santiago da Compostela, a beach paradise called Miami. At century's end, like abandoned offshoot shrines, derelict hovels that once served as shelter, refectory, and grange chapel for wayfarers on pilgrimage, now stand as secular relics in a profane world of crass consumerism and competition. There are new signs of the times along the Dixie Highway: motels with broken neon lights or signs saying "CLOSED FOR THE SEASON" in barely visible letters and hanging by a single nail and a spider web; mom-and-pop lunch diners and dinettes, displaced by highway, as well as gastronomic realignments, and now serving only local townsfolk. Rusty gas pumps on islands fronting ceramic-clad structures whose unswept drive, broken corner glass, and sign advertising gas for 23.9 cents are tell-tale signs of societal detour. What began as a pilgrimage route for holiday makers and secular sun worshipers has become a pilgrimage route for historians, preservationists, popular culturalists, and converts of dissimilar denomination. At the wailing wall, there is no discrimination, only shared songs of lamentation.

How far the reader may be willing to take this extended metaphor may be unpredictable. One must not forget that the SCA was founded on pilgrimages to New England diner sites. As born-again crusaders, our founding members, those ancestral Deco-philes of what I have nicknamed the Streamlined Church of the Automobile (SCA), have proselytized and converted and gathered new initiates over the years; they have educated novices in the rites and catechisms of the road, and they have born witness to a growth of followers to rival Cluny in tenth- and eleventh-century Burgundy.

As the end-of-the-millennium year 1000 approached long ago, our brethren afoot looked back over the dark ages to a Roman era when engineers had been the great road builders across Europe; creating a neo-Roman style called Romanesque, the millennium generation became the great builders of monastic shrines and abbey churches at unprecedented scale and numbers. Approaching another millennium, roadside architects looked back to an era of the recent past, when the automobile stimulated a great era of U.S. road building and the aesthetic of the highway, which I will refrain from calling "roamin' architecture," has now stimulated another revival, the neo-Moderne. Out of the Deep South, in the shadow of Carl Fisher's mecca, the CEO of the Checkers company, Jim Mattei, devised a concept in 1985 that became the Checkers Drive-In, with prototypes built in 1989–90. From Florida, like sister houses of Cluny, Checkers outlets have proliferated across the national landscape. Dixie diner Deco has been born again.

And there is an international missionary field! During the past ten years, initiates have gathered at World Congresses on Art Deco, international gatherings of the devout that neither Carl Fisher nor Norman Bel Geddes could have envisioned. The first World Congress convened in the South Beach Historic District of Miami Beach, where Santa Barbara has since been canonized, and revival meetings have been held biennially ever since, including recently on the *camino real,* Route 66 in middle America.

Despite the Dark Ages of a Dixie Highway being erased from the landscape, new shrines of the road have been sanctified and Deco devotees baptized. Like wayfarers of old, modern-day SCA pilgrims, seeking Santiago, continue to find glass block and neon, streamlining and lines of speed, in Checkers and even Deco-ized McDonalds sited along the highways of Dixie. The story of the Dixie Highway is a tale worthy of Chaucer. Lined up in a row of counter barstools at the Hot Rod Café, one should not be surprised to find the modern-day reincarnations of the [NASCAR] Knight and Squire, a Yeoman and Merchant, a Clerk, a Sergeant of the Lawe, a Doctour of Physik, and the Wife of Bath.

. . . And after wol I telle of our viage,
And al the remenant of oure pilgrimage.

. . . Also I praye you to foryive it me
Al have I nat set folk in hir degree
Here in this tale as that they sholde stonde:
My wit is short, you may wel understonde.[22]

NOTES

1. The first "See Rock City" barn was on the Dixie Highway north of Chattanooga. Howard Lawrence Preston, *Dirt Roads to Dixie: Accessibility and Modernization in the South, 1885–1935* (Knoxville: U of Tennessee P, 1991), 152. Cited in Martha Carver, "Dixie Highway Bus Tour Two: Moon Pies, Monuments, and Motels," in Jeffrey L. Durbin and Martha Carver, *Drivin' the Dixie: Automobile Tourism in the South* ([Nashville, TN]: Society for Commercial Archeology, 1998), section E, p. 39.
2. The *Liber Sancti Jacobi,* or Pilgrim's Guide, dating from between 1139 and 1165, is attributed to Aimery Picaud, born in the village of Parthenay le Vieux, where the twelfth-century Church of St. Pierre survives. It is a Latin manuscript (of which several copies survive in Compostela, Paris, Rome, London, Ripoll, and Alcobaça, Portugal) that is also known as the *Codex Calixtinus* because Calixtus II was wrongly considered its author. Its five books comprise a series of sermons and hymns; an anthology of miracles performed by St. James; a collection of stories associated with the life of St. James and the discover of his tomb; the *Historia Karoli Magni et Rotholandi,* primarily treating the history of Compostela and the campaigns of Charlemagne; and a medieval Pilgrim's Guide. The guide "sketches out the French and Spanish routes, points out the chief shrines, and offers advise to the traveller on all sorts of topics." Brian and Marcus Tate, *The Pilgrim Route to Santiago* (Oxford: Phaidon, 1987), 17, 20.
3. Geoffrey Chaucer, "The General Prologue," in *The Canterbury Tales,* ed. M. H. Abrams et al., *The Norton Anthology of English Literature,* vol. 1 (New York: W. W. Norton, 1962), 99, lines 696–702.
4. Margot Ammidown, "Edens, Underworlds, and Shrines: Florida's Small Tourist Attractions," *Journal of Decorative and Propaganda Arts* 23 (1998): 249. (This was the journal's Florida theme issue.)
5. Hap Hatton, *Tropical Splendor: An Architectural History of Florida* (New York: Alfred A. Knopf, 1987), 184.

6. Geoffrey Chaucer, "The General Prologue," 99, lines 703–8.
7. "Edward Bok and Bok Tower," in Hatton, *Tropical Splendor,* 163.
8. Gonzalo de Berceo, *Los milagros de Nuestra Señora,* ed. Brian Dutton (London, 1971), 29, quoted in Tate, *The Pilgrim Route to Santiago,* 13.
9. Weeki Wachee Springs, north of Tampa, was off the Dixie Highway but comparable.
10. See "Aquatecture: The Steamboat in Florida" in Hatton, *Tropical Splendor,* 17–19.
11. The *African Queen* itself is in Key Largo, Florida. A thirty-foot open hulled steam launch built circa 1912 in England with galvanized steel hull had been employed on the rivers of Central Africa and was used in the filming of *The African Queen* (1951), starring Katherine Hepburn and Humphrey Bogart. The boat was placed on the National Register in 1992. Morton D. Winsberg, *Florida's History through Its Places: Properties in the National Register of Historic Places* (Gainesville: Bureau of Historic Preservation, Division of Historical Resources, Florida Department of State, distributed through UP of Florida, 1995), 81.
12. This was the description of Daniel Brinton, an early visitor to Silver Springs and an anthropologist and archaeologist. Richard A. Martin, *Eternal Springs: Man's 10,000 Years of History at Florida's Silver Springs* (St. Petersburg, FL: Great Outdoors Publishing, 1966), 107, quoted in Ammidown, "Edens, Underworlds, and Shrines," 245.
13. Ammidown, "Edens, Underworlds, and Shrines," 245.
14. Ibid., 252. Orchid Jungle opened in 1925 and Monkey Jungle in the early 1930s, both near Homestead, and these were followed by Parrot Jungle in Miami, which has received Dixie Highway travelers since the attraction opened in 1936 and remains popular.
15. John Margolis illustrates Gatorland Zoo, in Christmas, Florida, where "Swampy the Giant" alligator, half building and half sign, was erected as recently as 1989 to keep alive alligator wrestling and reptile souvenir collecting. Another Gatorland was built near St. Augustine. John Margolis and Emily Gwathmey, *Signs of Our Time* (New York: Abbeville, 1993), 74–75.
16. Ammidown, "Edens, Underworlds, and Shrines," 257.
17. Warren James Belasco, *Americans on the Road: From Autocamp to Motel, 1910–1945* (Cambridge: MIT P, 1979), 89. When the author's family traveled from St. Louis to Miami for mid-winter sojourns during the 1950s, the over-1200-mile journey (parts of it over the Dixie Highway) took three days. Michael Jacobs notes that a pilgrim walking from the Pyrenees to Santiago could average at best 15 miles per day. Michael Jacobs, *The Road to Santiago de Compostela* (London: Penguin Books, 1992), x.
18. John Ruskin, "The Lamp of Memory," in *The Seven Lamps of Architecture* (New York: Noonday, 1961), 176–77.
19. Ruskin wrote, "Great nations write their autobiographies in three manuscripts;—the books of their deeds, the book of their words, and the book of their art. Not one of these books can be understood unless we read the two others; but of the three, the only quite trustworthy one is the last." John Ruskin, preface to *St. Mark's Rest,* quoted in Kenneth Clark, *Ruskin Today* (Harmondsworth, UK: Penguin Books, 1964), 180.
20. Not all members of the cloister endorse the pseudonym of the SCA, the Streamlined Church of Automobiles. It is in recognition that Rev. Robert Schuller of Crystal Cathe-

dral fame got his start preaching in a drive-in movie theater in Los Angeles that the religious metaphors herein contained, finding in car culture a religious experience, may be deemed appropriate.

21. Examples of applying the word *Dixie* to roadside architecture along the Dixie Highway include C. S. Little's Dixie Camp, located on the Eastern Division (U.S. 1) of the Dixie Highway, south of Waycross, Georgia, and the Dixie Motel north of Ringgold (2.7 miles south of the TN/GA state line). Jeffrey Durbin cites other examples in Georgia: "Donald C Brock's Camp Dixie (a combination restaurant, gas station, and cabin camp) near Tunnel Hill; Dixie Car Wash and Dixie Auto Wash in Dalton . . . Dixie Cab in the Atlanta suburb of Forest Park; Augusta's Dixie Court; Vernon Smith's Dixie Motel in Adel; Valdosta's Dixie Court; the Dixie Pine Motor Court in Bainbridge; and the chain of Dixie gas stations still found in Georgia towns such as Calhoun and Baxley." Jeffrey L. Durbin, "The Dixie Highway in Georgia" in Durbin and Carver, *Drivin' the Dixie,* section B, p. 37.
22. Geoffrey Chaucer, "General Prologue," 100, lines 725–26, 745–48.

CONTRIBUTORS

MARTHA CARVER has been a historic preservation specialist and historian with the Tennessee Department of Transportation (TDOT) since 1980 and has supervised TDOT's historic section since 1984. She has a bachelor of arts degree in history and a master of arts with an emphasis in historic preservation. She has written two previous articles on the Dixie Highway for the *SCA Journal* as well as an article on folk artist Harrison Mayes, has written TDOT's historic bridge survey, and was a co-author for the SCA 1998's conference tour book on automobile tourism in the South. She has also written entries for the *Tennessee Encyclopedia of History and Culture* and the *Encyclopedia of Appalachia*. Professional interests include road and bridge history as well as roadside architecture. She collects historic postcards.

ROBERT M. CRAIG is a professor in the College of Architecture, Georgia Institute of Technology, with a specialty in architectural history. He served two terms on the National Register Review Board for Georgia, completed six years on the Atlanta Urban Design Commission, including a term as commission chair, and is a charter member of the Atlanta Preservation Center. Craig was a founding member and past president of the Southeast Chapter, Society of Architectural Historians, and was elected to the Board of Directors of the national Society of Architectural Historians and serves as their secretary. In 1999–2000 he was president of the Southeast American Society for Eighteenth-Century Studies and is currently president of the Nineteenth Century Studies Association. From 1983 to 1987 Craig was editor of the *SECAC Review* (the scholarly art journal of the Southeastern College Art Conference), and he continues to serve as a member of its editorial advisory board. He has authored books on Atlanta architecture and is a frequent contributor of essays on nineteenth- and twentieth-century architectural topics in international encyclopedias and dictionaries of architecture. Craig has lectured and presented scholarly papers in China, Australia, Canada, France, and throughout the United States. His special interests include roadside architecture, the Arts and Crafts movement, nineteenth- and twentieth-century architecture, and Chinese gardens. His most recent book is *Bernard Maybeck at Principia College: The Art and Craft of Building* (2004).

JEFFREY L. DURBIN is a historic preservation specialist with the Advisory Council on Historic Preservation, an independent federal agency that advises the president and Congress on historic preservation issues. He previously worked with URS Corporation in Gaithersburg, Maryland, and the Georgia State Historic Preservation Office in Atlanta. His entry about the Boone Tavern in Berea, Kentucky, appears in the *Encyclopedia of Appalachia*. Along with Lisa Raflo, he cowrote "Teal Roofs and Pecan Logs: A History of Stuckey's Pecan Shoppes," which appeared in the *SCA Journal*. Durbin is a former vice president of the SCA and a current board member for the Pioneer America Society.

SUZANNE FISCHER holds a bachelors of science in geography from Michigan State University and a masters of arts in public history from Indiana University/Purdue University at Indianapolis. She prepared a historic context study, "Facilities for Motorists 1900–1940," which served as the basis for her master's thesis on the Dixie Highway in Indiana, from which her chapter was adapted. Fischer served on the SCA board 1999–2006 and hosted the organization's conference in Indianapolis. She has worked as an architectural historian for a cultural resource consulting firm in New Orleans and is currently the curator of material culture for the Loui-siana State Museum.

KATHERINE FULLER received her masters in historic preservation from Goucher College in 2002. Her thesis, "Populuxe Motels: Preservation in Myrtle Beach, South Carolina and the Wildwoods, New Jersey," examined the economic challenges of preserving these two resort communities.

KIMBERLY D. HINDER is the senior preservation planner with the City of St. Petersburg, Florida. Prior to that, she was the senior architectural historian with Archaeological Consultants in Sarasota, Florida. She has served as an archivist, a public records manager, and a research historian for several consulting firms and government agencies in both Florida and Georgia. She has a master's of historic preservation from the University of Georgia and a bachelor of arts in history from Agnes Scott College in Atlanta. She has published on historic tourist attractions and supermarkets. Her current research interests include the evolution of racing at Daytona Beach, space-related roadside architecture, and mid-century modern residences.

ROBBIE D. JONES is the director of preservation and editor of *The Jacksonian* at the Hermitage, Home of President Andrew Jackson, in Nashville, Tennessee. He holds a bachelor of architecture from the University of Tennessee and a master's in history from Middle Tennessee State University. He wrote *The Historic Architecture of Sevier County, Tennessee,* and he contributed to *Coastal Plain and Fancy: The Historic Architecture of Lenoir County and Kinston, North Carolina*. Jones recently completed a study of progressive-era civic reform and the development of Carnegie libraries in the South.

RUTH NICHOLS KEENOY is a historic preservation consultant. Keenoy completed a master of arts in history from Middle Tennessee State University in 1994. Her publications include several articles in the *Tennessee Encyclopedia of History and Culture* (1998) that are related to Civilian Conservation Corps projects in Tennessee state parks, and also for the *Encyclopedia of Appalachia* (2006) related to New Deal–era projects in Tennessee and West Virginia. Keenoy resides in St. Louis, Missouri, with her husband, Patrick, and stepson, Jack.

SARA AMY LEACH is senior historian at the National Cemetery Administration, Department of Veterans Affairs. Previously Leach spent thirteen years with the National Park Service (NPS), Department of the Interior, working as a historian and cultural resource manager. This experience includes five years at the Natchez Trace Parkway in Mississippi, first as the chief of interpretation and visitor services and subsequently as the park's cultural resource specialist. Prior to that she was a project historian with the Historic American Buildings Survey (HABS), NPS. She has a master of arts in architectural history and a certificate in historic preservation from the University of Virginia, as well as a bachelor of arts in journalism and bachelor of fine arts in art history from Ohio Wesleyan University. Among her research interests are commercial roadside buildings, historic highways/parkways, urban planning, and cemeteries.

CHRISTIE H. McLAREN was the special projects historian at the Department of Arkansas Heritage. She has researched Arkansas highway history leading to the completion of the National Register of Historic Places context "Arkansas Highway History from 1913–1965." She is an archivist and has received fellowships from the Library of Congress and the American Philosophical Society Library of Philadelphia. She completed her history and archival science studies at Auburn University. A native of Southern Missouri and Arkansas, she now resides in Arkansas with her family.

WALTER S. MARDER is a historic structures consultant in Florida. He was a preservation architect at the Florida Bureau of Historic Preservation until May 2005. Prior to working for the State of Florida he was the architect for the city of Wilmington, Delaware, and the acting chief architect for the Ministry of Works, U.S. Peace Corps, in Belize. He has a bachelor of architecture from Rensselaer Polytechnic Institute and is working on a master of history degree at Florida State University. Marder is a registered architect in Florida and California. He has contributed to articles on historic buildings and materials for several publications.

KEVIN J. PATRICK peered out the window while on family trips to the Jersey Shore or to visit relatives in western Pennsylvania and thus built up a mountain of questions that could only be answered by acquiring degrees in geography: a bachelors from New Jersey's Glassboro State College in 1983, a masters from the University of Illinois in 1985, and a doctorate from the University of North Carolina in 1996. His dissertation, "Learning from Lincoln Highway: Identity, Place and a Pennsylvania Roadscape," began a long-term commitment to the interpretation and heritage planning of historic highways that has yet to abate. He was the editor of the national quarterly *Lincoln Highway Forum* from 1997 through 1999, and he has done consulting work on four different Lincoln Highway projects, including the recent Lincoln Highway Special Resource Study undertaken by the National Park Service. His road- and roadside-related publications include "The American Roadside Giant: Myth, Meaning and Cultural Identity" in the spring-summer 1995 *SCA Journal; Diners of Pennsylvania,* a book he co-authored with Brian Butko in 1999; and the book *Pennsylvania Caves and Other Rocky Roadside Wonders,* published in 2004. Kevin Patrick has been teaching geography at Indiana University of Pennsylvania since 1993.

KARL PULJAK is an associate professor and interim director of the School of Architecture at Louisiana Tech University. He has a bachelor of architecture from Kansas State University and a master of architecture from Cranbrook Academy of Art in Michigan. He is also coprincipal of Puljak or Willoughby, Design, of Ruston, Louisiana.

CARRIE SCUPHOLM is currently pursuing her doctorate in architectural history and urban planning at the University of Florida. She received her master of arts in historic preservation planning from Cornell University and her bachelor of arts in art history from the University of Michigan. Prior to returning to school, she spent ten years working as a historic preservation consultant for private consulting firms, nonprofit organizations, and state agencies. Her research interests include women in architecture, segregation and recreational travel in Florida, and the roadside architecture of Florida's early tourism industry. Scupholm is in her third term as an SCA board member and serves as the organization's president.

BLYTHE SEMMER, is a historic preservation specialist with the Advisory Council for Historic Preservation. She studied the Smokies' roadside architecture for her master's thesis at Middle Tennessee State University. An East Tennessee native, she holds a bachelor's degree in American studies from the College of William and Mary. Her preservation background includes researching highway bridges for the Historic American Engineering Record, working as a historian for the Alabama Historical Commission, and working as a preservation planner for Nashville's Metropolitan Historical Commission.

R. STEPHEN SENNOTT is an architectural historian and editor and the assistant dean for academic affairs and an adjunct associate professor in the College of Architecture at the Illinois Institute of Technology, where he teaches the history of architecture. His research on Chicago architecture, planning, and the automobile has appeared in Jan Jennings, ed., *Roadside America: The Automobile in Design and Culture* (1990), and John Zukowsky, ed., *Chicago Architecture and Design, 1923–1993: Reconfiguration of an American Metropolis* (1993), the *AIA Guide to Chicago* (1993), and *The American National Biography* (1999). He edited the three-volume *Encyclopedia of Twentieth-Century Architecture* (2004).

CLAUDETTE STAGER is a historic preservation specialist for the Tennessee Historical Commission and has been coordinating the National Register of Historic Places program for the state since 1985. She has a bachelor of arts from Syracuse University. After graduating from Eastern Michigan University with a master of science in historic preservation, she worked as a consultant in Wisconsin. Stager has been a presenter at various local, state, and national events and written articles on preservation topics. She is a former vice president of the SCA and has also served on the board of Nashville's historic City Cemetery.

INDEX

Page numbers in **boldface** refer to illustrations.